Education Division 2019

Held at the 2019 AIChE Annual Meeting

Orlando, Florida, USA
10 - 15 November 2019

ISBN: 978-1-7138-0528-1

Printed from e-media with permission by:

Curran Associates, Inc.
57 Morehouse Lane
Red Hook, NY 12571

Printed with permission by Curran Associates, Inc. (2020)

For permission requests, please contact AIChE
at the address below.

AIChE
120 Wall Street, FL 23
New York, NY 10005-4020

Phone: (800) 242-4363
Fax: (203) 775-5177

www.aiche.org

Additional copies of this publication are available from:

Curran Associates, Inc.
57 Morehouse Lane
Red Hook, NY 12571 USA
Phone: 845-758-0400
Fax: 845-758-2633
Email: curran@proceedings.com
Web: www.proceedings.com

TABLE OF CONTENTS

566102 You Can Do It! Overcome Common Challenges in Engineering Education Research

Are you interested in performing engineering education research studies but intimidated by the many challenges posed by experimental design, IRB approval, as well as choosing the appropriate quantitative and/or qualitative assessment methods for your work? Are you wondering how to expand your conference papers to more rigorous journal publications? If so, this workshop is for you! Two members of the Education Division will share stories of successes, oversights and lessons learned from their experiences completing research studies with the goal of generating discussion to help fellow researchers.

574818 Flow Chemistry Case Study for Material and Energy Balance Course

A case study for the undergraduate material and energy balance course was developed based on the 2017 publication of a flow chemistry process by Eli Lilly and Company. The case study is presented as homework problems that can be distributed throughout the semester, covering topics such as material balances on non-reactive and reactive, single-unit and multi-unit processes, liquid-liquid, vapor-liquid, solid-liquid and solid-vapor equilibrium, and energy balances on reactive and non-reactive processes.

567966 A Summer Overseas Experience at the Imperial College CO_2 Capture Pilot Plant

The Notre Dame administration strongly encourages departments and colleges to arrange programs for students to study outside of the US. It has an extensive organization to aid with this and owns/leases properties in about a dozen countries to support the effort. Our students embrace these opportunities with about 75% of ND students participating in some sort of international educational experience. For engineers, who have much less curricular flexibility, the participation rate is somewhat lower, about 60%, with most of these students attending 6-week summer programs in Rome, London, Alcoy Spain, Berlin, Dublin. Given the normal career progression for engineering students (almost all Notre Dame engineers graduate in 4 years), summer after Junior year is reserved for an internship, perhaps Sophomore as well, thus a great many of these students will have just finished just their first year at Notre Dame.

Five years ago we were presented with an opportunity to participate in summer courses at Imperial College London. The chemical engineering department at Imperial has what could be the only facility of its kind, a pilot-scale (50 kg/kr gas flow) carbon dioxide "capture" faculty. A nitrogen - carbon dioxide mixture is absorbed into a monoethanolamine solution in a packed tower (~7 m height), with steam regeneration in a second tower. The facility is fully-instrumented in a "catalog" sense in that different components are chosen (e.g., shell and tube, plate, spiral heat exchangers; vortex, Coriolis, swirl and Venturi flow meters, structured and dumped packing) to provide students with experience in the performance of many different types of components. Even the control systems have, *Profibus*, *Fieldbus*, 4-20 mA and some WiFi communications protocols.

Our program needed to occur after Sophomore year (first year is not possible). Students will have had the material and energy balance course, and the first thermodynamics course but no background in fluid flow, heat transfer, mass transfer, phase equilibrium or process control. All essential topics in understanding how the pilot plant operates. A similar problem exists for the companion course, a replacement for our Junior level experimental laboratory, which has a heavy emphasis on data analysis and report-writing. Almost identical academic limitations exist for the students as the normal semester for this lab course is (the 6th) spring of Junior year.

To meet these challenges, a preparation class is held one hr/week during the preceding spring semester. Lecture topics are organized around the *Unit Operations*, fluid flow, heat transfer, mass transfer, process control. The laboratory experiments are chosen to align with these, (1) fluid flow and measurement, (2) performance of different heat exchangers, (3) gas absorption in a small packed column and (4) Rankine cycle or vapor compression cycle. Modest amounts of homework is given focusing on getting the correct "numbers" for say, mass and heat transfer coefficients from simulated experimental and process data. An almost weekly part of the class is taken up discussing the workings of the CO2 absorption/desorption process based on the P&ID diagram — (e.g., what adjustments to flow rates or heat input would reduce the CO2 concentration in the exit gas? What temperatures are needed in the various devices?). The specific names of all of the components (e.g., H100, J101, C100, C400, etc.) — essential when students explore and operate the process — are used from start.

The talk will describe these various challenges (including report writing) and how we have adjusted the preparation course, and time at Imperial to meet them. An example of an attempt at the *most efficient way* to teach "mass transfer" will be given. While initially the student reports were below what was desired, the Pilot Plant grades from the Imperial College instructor have improved substantially with 2 (of the 4 submitted in the last 2 years — pilot plant groups are 12-14 students) being A's and other two just below the A/B cut-off. Likewise, the written reports for the laboratory course (which is graded by ND faculty) have reached a level that matches what our Juniors produce.

571855 Design Challenges As a Spine to Chemical Engineering Courses

Chemical engineers have led many innovations in engineering education, from publishing one of the very few long-standing discipline-specific engineering education journals, to hosting summer schools for new chemical engineering faculty, to promoting active, self-paced learning, such as through http://www.learncheme.com/. Yet many aspects of undergraduate education remain relatively unchanged. As one of three chemical engineering programs that received a National Science Foundation RED "Revolutionizing Engineering and Computer Science Departments" grant, we have taken our particular context as a Hispanic-serving, research institution and embraced it to address these concerns. To enhance opportunities for our diverse students and connect their previous life and cultural experiences to chemical engineering, we have developed design challenges that have a low-bar entry, high ceiling threshold, and provide concrete connections to core content. For instance, in the first year course, students investigate the 2015 spill of acid mine drainage in the Animas river. Students choose a rural community, propose a community engagement strategy, develop a prevention or emergency response water filtration system, and deliver an oral presentation of their designs. In one sophomore design challenge, students apply what they are learning about chemical process calculations to the design of an algal biofuel plant scaled for a specific rural community. Students use a decision matrix and conduct techno-economic modeling to choose ways to grow, harvest and extract algal fuel. The challenge culminates in a public blogpost that builds on the team wiki page posts. In a junior design challenge, students apply what they are learning about transport phenomena to investigate the Kirtland Air Force base jet fuel leak. Students use diffusion, advection, chemical kinetics, and separations to design mitigation and containment solutions. Analysis of student learning demonstrates that this approach does not come at a cost to conceptual knowledge gains, reveals more about student depth of understanding than traditional assessments, and provides opportunities for students to develop critical professional skills (e.g., teamwork, ethics, communication, and critical thinking). Student surveys and course assessments suggest that such an approach to enhancing engineering education may have a positive effect on the retention of students from demographics underrepresented in engineering, especially rural and first generation students.

576016 Development of the Chemical Engineering Identity from the Freshman to the Sophomore Year

The development of a STEM identity in students has been a topic that has received a growing amount of interest for more than a decade. The successful development of a STEM identity has been positively linked to success and persistence within a discipline and ultimately the attainment of a degree. While the majority of studies on identity development has treated STEM identity, particularly engineering identity, as common among all disciplines, more recent studies have begun to show that there are significant differences in the development of identity between the engineering disciplines. One of the well-studied student populations within engineering are those within Chemical Engineering where it has been initially shown that the development of a Chemical Engineering identity across different dimensions is not only different from that of other engineering disciplines, but is also different from that of chemistry. In this work, we will build on the idea of the development of the Chemical Engineering identity from the start of the student's freshman year through their completion of the first chemical engineering courses in the beginning of their sophomore year. We find that many Chemical Engineering students begin their academic careers with a somewhat developed sense of identity that is different from that of the general engineering student body. Additionally, certain aspects of their identity, including their career aspirations and ability to speak with content experts, develop quickly through the beginning of their Chemical Engineering curriculum.

573412 Can Enriched Extracurricular Activities Influence Early Undergraduate Students' Interest and Participation in Chemical Engineering Design Teams?

Conventional wisdom advocates that engineering student's involvement in extracurricular activities is one of the key factors for a fulfilling and beneficial college experience. Engineering literature shows that extracurricular activities such as professional organization and participation in design teams can positively impact undergraduate students, in particular, those from underrepresented communities. The current work explores the use of extracurricular professional development workshops to motivate and engage early undergraduate students on enriched process design activities. Student primary motivators for participation are building their strength for future job hiring, building community with their peers, and to represent our institution during the design team competition hosted by AIChE every year.

The first cohort of the professional development workshops invited students to engage voluntarily in chemical engineering design skills. Only students that were in their first, sophomore and junior years were accepted to participate. Students online instructional videos independently and then partake in design challenges in a collaborative setting. Video lessons focused on completing simulation exercises that introduce common industry equipment (e.g., Reactors, Pumps, Heaters, Coolers, Mixers, and Tees), and processes. Modeling and simulation of these processes were facilitated by using UniSim/HYSYS tools. Each face-to-face workshop required students to work on team projects challenges using what they learned in the online tutorials to solve exercises at a higher level. The study will use an already validated pre- and post- questionnaire and open ended-questions to examine how attitudes before and after being impacted by the professional development workshops. The current work will present valuable information for designing enriched extracurricular activities that can potentially be used for curriculum improvement and participation in design teams.

566303 Measuring Spreadsheet Skills Using Analytics within an Interactive Textbook

Spreadsheets are a daily part of more chemical engineers' lives. Spreadsheets are a mature product and available in many forms, including Microsoft Excel, Google Sheets, and many others. Across versions, devices, and platforms (e.g., Windows, Mac, iOS, Android), many formulas, functions, and tasks are common. Spreadsheets Essentials is an interactive textbook from zyBooks that allows students many opportunities to click and learn new material. Examples include stepping through animations, completing matching question sets, and perform spreadsheet functions on auto-graded, randomized problems. Beyond basic spreadsheet usage, advanced topics are available, such as solver, interpolation, numerical integration, error, statistics, and matrix operations. Interactivity creates big data aggregated in real time for instructors via student clicks and data entry. Using a small fraction of the final course grade as an incentive, reading rates above 90% were observed in 2018 and 2019. Success on over 120 auto-graded questions will also be included. Finally, determining which auto-graded questions take students the most attempts before correct is one question to be addressed using big data created by students using an interactive textbook.

575936 Retention Improvement Efforts in the Undergraduate Living and Learning Community at the University of South Carolina

The University of South Carolina Living and Learning Community works to help students succeed in their academic careers. Current initiatives include a professional development course, linked mathematics courses, and local tutoring availability. New initiatives to support student retention will be discussed. These efforts include:

1. Boot camp - A short early move-in option helps students adjust to campus and form social networks while providing mathematics review and professional development information.

2. Additional linked courses - In addition to mathematics, the living and learning community now has implemented linked humanities coursework so that students living together can take courses with other residents.

3. Spring activities - Traditionally, most efforts have focused on acclimation in the Fall semester. New efforts help support students throughout their first year.

568066 Steal This Elective: Applied Food Science and Engineering

This upper-level elective is designed to introduce chemical engineers and other STEM-undergraduates to the application of their disciplines to food processing and food science through a problem-based-learning approach. Each "problem" is a real-world food issue, carefully selected such that all students have engaged with all course content outcomes by the end of the semester. Past PBL units include: reformulating tomato sauce as a result of a poor tomato crop; development of new ice cream flavors for a medium-sized business; process design for 'veggie burger' production, and addressing the challenges of local food deserts. Students learn the basic chemical composition of foods, common reactions, approaches to food preservation, and how codes, standards, and regulations apply to food production and sales. The course also cultivates students' broader outlook and professional skills through deliberate incorporation of food's ethical, cultural, and societal dimensions. The course is structured so that laboratory and lecture periods are flexibly interchangeable, allowing for a wide variety of activities during any given class period.

574331 CRISPR in the Curriculum: A Genetic Engineering Elective for Chemical and Biomedical Engineers

The last 50 years have seen major strides in gene editing technology. Not only is this topic currently popular, but also increasingly relevant to society. As our students navigate decisions around genetically modified organisms in the food supply and genetic engineering for medical applications, a basic understanding of the technologies, their limitations, and potential impacts becomes almost a necessity. This talk will describe a relatively new genetic engineering elective designed for Chemical and Biomedical Engineering students. The course is designed to accommodate a breadth of biology backgrounds--students range from having last seen biology in high school to having several semesters of post-secondary biological courses--and focuses on understanding and assessing the technology in context. Topics within each of these modules are introduced through the lens of historic or current societal issues, with emphasis on defining the full context of the issue, including but not limited to the social, economic, political, geographic, religious, and historical contexts. The course runs in three modules: (1) basic biology and genetics refresher, (2) genetic engineering technologies, and (3) current applications and advances in genetic engineering. The term project challenges students to dive deeper into a societal issue of their choice, and ultimately propose and assess a genetic engineering solution to the issue. Students present the synthesis of their work through an educational board game, designed to teach players about their societal issue and technology, and equipping the students with the ability to engage in these topics with a broad range of audiences in the future.

579479 Process Prototype Design and Fabrication

Quite often chemical engineering students will get to test the fundamentals of what they learn in the classroom in the laboratory such as unit operations, fluid and heat transfer, and reaction kinetics. However, it is rare that they get the chance to design and fabricate prototype systems. Indeed, allowing students to design and build laboratory scale chemical engineering hardware can enable creativity, resourcefulness, and awareness of the intricacies of how hardware is assembled and operates. Fundamentally, it allows students to engage in an open-ended creative process and build awareness of process safety. The class starts with a set of needs for equipment in the Unit Operations Lab or a research group. Each student prepares a sketch and makes a pitch of a possible solution to the need. At each stage in the design process, the potential solutions are reduced until only one remains the project teams are formed to refine the design and build the hardware. The final deliverable is a demonstration of the hardware with an operations manual. This elective course has been offered twice at Kettering University and while it is capital intensive, the resulting hardware is durable enough to be used in the teaching laboratory. In fact, the course is less expensive than purchasing teaching hardware from vendors. Some hardware is remains functional after five years. Fabricated systems include fermenters, fluidized beds, plug flow reactors, and Labview data acquisition systems. Examples of the design process, the course, and developed hardware will be presented.

559088 Building Computational Skills for Mathematical Modeling in Science and Engineering through an Interdisciplinary Elective Course

An upper division elective course for seniors and graduate students in science, technology, engineering, and mathematics (STEM) fields has been developed to build computational skills for mathematical modeling. The course aims to fill a gap in the practical training of students starting computational research projects across various STEM disciplines who have inconsistent previous experiences in computer programming and numerical methods. By accelerating the training time to develop competency in implementing modern best practices, students are enabled to be productive at using computational tools for research early in their graduate studies, ideally allowing them to satisfy time-sensitive demands for generating research results. Undergraduate students can also benefit from learning practical computational skills that they can transfer into other courses and future projects in industry or other career paths. The practical training is achieved by covering modern software tools for mathematical modeling in science and engineering and for reproducible research computing via an active, hands-on approach supplemented by reading materials. Rather than covering just the basics of programming or detailed algorithms for numerical methods, the course is geared towards implementing tools for solving realistic continuum scale science and engineering problems, managing open source code projects, and disseminating computational research results through scientific documentation and publications. The course is taught by a chemical engineering faculty member with research expertise in applied mathematics and computational science and engineering. MATLAB and Python are taught side-by-side throughout the course, primarily using case studies from chemical engineering applications. Additional modern software tools including Git and LaTeX are also covered in the course. The presentation will describe the course with the goal of enabling other educators to adapt and reuse the course content.

559710 Enhancing Global Understanding of Sustainable Energy Needs By Running a Faculty-Led International Program

Sustainability is of global concern with the increasing awareness of humans' impact on the environment. Teaching engineers about global sustainability issues and how to produce innovate solutions to these problems can be challenging in a traditional classroom. Leading an international faculty-led program can enhance student learning of important global problems and help drive innovation to solving these critical issues. There are added benefits to students having an international experience by becoming globally aware and global citizens, more personally and socially aware, and strengthening their career and academic choices. This talk will discuss the lessons learned running a successful one-month long faculty-led study abroad program to Brazil seven times.

As part of this one-month program, the students took two second year level courses. *Energy Systems* which discusses the need for sustainable energies and the scientific and engineering principles on how they work and *Impacts of Alternative Energy Technologies* which discusses the need to understand environmental and social justice issues related to energy systems. During the program, the students visited multiple renewable energy sites and talked with industrial leaders who pioneered these technologies in Brazil. They are also immersed in Brazilian culture while learning and discussing innovative sustainable energy technologies by developing a sustainable urban development proposal for the city of Fortaleza with Brazilian university students from UNIFOR and ICLEI – Local Governments for Sustainability.

Topics discussed during this talk include managing student expectations and preparation before going abroad, impactful service learning projects, and managing working with in-country partners. Anecdotal evidence from students' blogs about their journey will be highlighted to demonstrate learning outcomes and experiences during the global experience.

574262 Scientific Computing for Chemical Engineers in Python

One of the most important "tools for modern engineering practice" is the ability to write computer programs. It is increasingly essential that our students learn how to code, both for their potential for future chemical engineering practice or to give them the skills they need to pursue a job in developing software.

In my required junior-level course, students learn numerical methods by applying them to chemical engineering problems in material and energy balances, thermodynamics, fluid flow, heat transfer, separations, and chemical reactor analysis. Topics include floating-point calculations and round-off error, algorithms and convergence, finding roots by bisection or Newton's method, exact curve fitting, interpolation, extrapolation, numerical integration and differentiation, numerical solution of initial value problems, stiffness, matrices and determinants, matrix properties, special matrices, methods of solution for systems of linear equations using matrices, eigenvalues, eigenvectors, solving systems of non-linear equations, and their applications to unit operations.

The course was previously taught in Matlab but is now in Python for the Spring 2019 semester. I will discuss the content of the course, some example problems, the difficulties in transitioning the course from one language to another, and preliminary results on year-over-year student performance.

558403 Hands-on Training on Intensified Membrane Processes for Wastewater Treatment

Hands-on training on intensified membrane processes for wastewater treatment

Itzel Marquez, Eduardo Saez, Kimberly Ogden, Andrea Achilli

ABSTRACT

Since the 1960s membrane processes have gained an important place in the chemical industry. The key properties of membranes that allow the selective permeation of a chemical species through the membrane has been exploited by processes such as liquid and gas separation and drug delivery. In the mid-1990s, thanks to the development of reliable and economical microfiltration and ultrafiltration systems, the water and wastewater treatment areas have become one of the most rapidly growing areas of membrane technology. Currently, membrane systems are competitive with conventional water and wastewater treatment producing treated water of superior quality (1). More recently, as water scarcity is becoming a persistent issue, water desalination and wastewater reuse have been pursued as alternatives to depleting water sources. Membrane technology provides efficient and effective means of treatment for several water quality feeds, using conventional as well as intensified processes to produce potable water from impaired sources.

Despite the fact that membrane processes and intensified processes can contribute to the solution of water scarcity, there is a lack of training resources and awareness in Process Intensification (PI) and advanced membrane processes. The American Institute of Chemical Engineers, through the RAPID initiative addresses this deficiency by creating PI and MCPI awareness by augmenting educational content introducing PI and MCPI topics relevant to the industry. As part of these efforts, this project is developing a hands-on course for graduate students and professionals to introduce the use of conventional membrane processes (ultrafiltration, UF, reverse osmosis, RO) and intensified membrane processes (membrane distillation, MD) to treat wastewater for potable reuse purposes.

The course will take place in the Water and Energy Sustainable Technology (WEST) Center at the University of Arizona where participants will have access to laboratories and controlled pilot/demonstration testing space for water and energy technologies. WEST houses bench- and pilot-scale membrane systems (UF, RO, MD) and has access to a wide range of water qualities (reclaimed, potable, and wastewater) onsite.

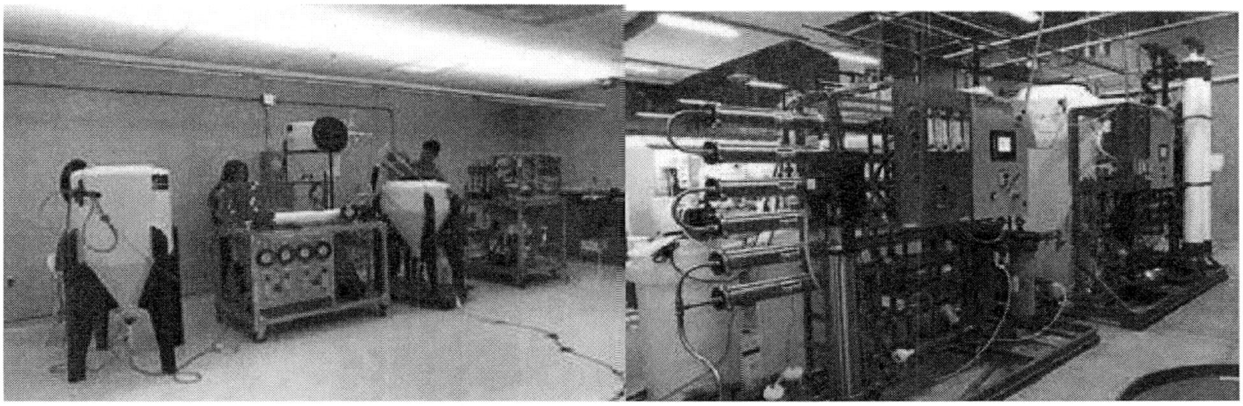

Figure 1. WEST center pilot scale systems

The participants will begin the course with an introduction to PI, the fundamentals of membrane processes and membrane process modeling software. Then, the participants will start proposing a treatment process train to remove contaminants from a given wastewater containing mixtures of particles, metals and/or inorganics. They will design and perform experiments testing their hypotheses in bench-scale and pilot-scale systems (UF, RO, MD). The experimental data will be used to calibrate process software developed by Chemstations, as well as software developed by our research group. Finally, they will perform economic and energy optimization of the proposed processes to establish practical viability.

By the end of the course, the participants will be able to:

1. Understand the operating principles of membrane processes (UF, NF, RO, and all varieties of MD) and their application in water treatment problems.
2. Operate bench- and pilot-scale treatment processes for water treatment.
3. Apply modelling software to perform economic and energy optimization of membrane processes.
4. Evaluate results from bench- and pilot-scale experiments
5. Design full-scale membrane treatment processes for the treatment of waters of impaired quality.
6. Understand the principles of Process Intensification in Membrane Distillation by doing experiments and modelling an MD full scale system.

Reference

1. **Baker, R. W.** *Membrane Technology and Applications.* s.l. : John Wiley and Sons Ltd, 2012.

579814 Training All Chemical Engineers in Computing and Data Science

In this contribution, I will discuss ongoing efforts to retool the sophomore-level "Numerical and Statistical Analysis" course (required) to provide a scaffolding for all students to develop core competencies in computing, applied statistics, and mathematical modeling throughout their undergraduate experience and profession careers. Beginning in Spring 2019, we are transitioning from MATLAB to Python for several reasons including consistency with "Chemical Process Control" (junior, required) and college-wide electives in data science and statistical computing that already use Python. I will also share experiences using Jupyter notebooks and cloud-based computing platforms such as Colaboratory to incorporate active learning into lectures and tutorials and to remove technical barriers for students. Content and assignments have been reorganized to emphasize mastery of foundational skills in preference over content breadth. For example, students are now required to submit hand-written pseudocode for all assignments to reinforce essential programming habits. We now emphasize *statistical reasoning* over more procedural approaches to data analysis, which required adding lectures on probability theory, a new topic for most students. In ten optional "extension assignments", students watch/listen to recent TED talks, podcasts, etc. focused on statistics, big data, and computing to build a broad appreciation for concepts introduced in the course. (As an incentive, students earn a homework "drop" by completing these assignments.) By developing fundamental skills, we seek to better position students to continue learning about numerical methods, data analysis, or scientific computing, either through technical electives at Notre Dame or other means as practicing professionals.

As part of the retooling process, lectures are being organized into 5 to 15-minute segments (instruction, activities, etc.) and will be recorded in front of a live classroom in Fall 2019. These 75-minute lecture recording will then be processed into short videos to flip the classroom starting in Spring 2020, thus allowing the two weekly 75-minute meetings to become structured problem-solving sessions.

581026 Digital Twin, Raising It out of the Realm of a Cliché

The Engineering and Industrial world are lagging manufacturing industries to truly embrace digitalization. AVEVA is accelerating the digital transformation journey of their customers and disrupting their current business processes to add value opportunities to their businesses. A common disruptive technology is the concept of Digital Twin. Through our work with our customers, we realize that believe to define a Digital Twin for the industrials we should draw a parallel to a the actual Human Twin. A twin has a shape, behavior, senses and needs to work for 100 years. The same is true for the digital twin of an industrial asset whether that is a an offshore platform, a refinery, a power generator or a ship. The Digital Twin of a plant is defined by engineering data (digital shape), engineering models (digital behavior), real-time series data (digital senses) and finally maintenance data (digital health). Once this data is connected just like a human it is can be stored in memory (historian), analyzed to make the right decisions (predictive analytics), learns from its mistakes (machine learning), asks questions (cognitive thinking) to make it more efficient (APC, real-time optimization), be mobile (smart devices) and finally visualize data in context (3D, X Reality etc...). AVEVA will present how it is connecting engineering, maintenance and operations data, people and business processes to bring the digital twin out of the realm of a Cliché.

580756 Asset Optimization Software Suites to Drive Digitalization – from Process Modeling to Optimization to Asset Optimization

Process Simulation has undergone significant transformation over the past decades driven by innovations in engineering best practices catalyzed by chemical engineering, mathematical optimization and software innovations. It has come a long way from automating the design calculations and process modeling to an environment for process optimization accelerating collaboration across the disciplines and tackling optimization across multiple dimensions simultaneously. The use of Process Simulation has also expanded from the traditional design applications to operations and planning. This digital transformation has created significant value in the industry resulting in capital and energy savings combined with better yields and capacities, increased safety and dramatic improvements in engineering productivity and production. Rapid advances in machine learning, advanced multivariate analytics, high performance mobile computing, and equipment and sensor connectivity are playing a role in enabling a whole new generation of technological solutions. The paper will cover the vision for Process Simulation including expansion from Process to Asset Optimization with first principles-based models and data-based machine learning models for process engineering, operations and performance management.

580855 A General Digital Applications Platform

Deep process knowledge captured within physics-based models is a key element for the successful digitalization of process industries. Digital applications making use of models derived from first principles are increasingly being used in industrial practice, particularly in the area of process operations. They include both open-loop applications, such as model-based soft sensing and monitoring of equipment degradation, and closed-loop ones such as Real-Time Optimization and nonlinear Model Predictive Control.

However, most non-trivial digital applications involve much more than a mathematical model being solved in a simulation or optimization mode. They often require multiple model-based calculations being scheduled over time and exchanging data with each other, while potentially being subject to occasional failures. They also involve extensive communication with external data servers, such as distributed control systems, plant historians, commercial databases and user dashboards. Moreover, the data involved in all such communications may be subject to systematic and/or random errors, and may even occasionally become unavailable.

This paper describes a recently developed general software platform for resilient and sustainable digital applications, taking account of the above considerations. The platform significantly reduces the cost and increases the reliability of development, testing, deployment and maintenance of diverse applications within a unified software architecture. Industrial examples illustrating the flexibility of the proposed design are presented.

580649 Concept of Digital Twin and Digital Thread in the Chemical & Process Industry

The vision of digital twins involves the digital twin of the product, the digital twin of process performance and the digital twin of operation.

In the R &D, stages product development and recipe transformation in with digital models can accelerate innovation. In design & engineering stages, all the disciplines involved in a plant project can generate a shared data model –saving valuable time and expenses and at the same time increasing engineering quality. The digital twin also allows the plant to be simulated prior to the critical commissioning phase, which ensures a reliable implementation. The operators can also receive targeted training before the system is commissioned. The digital twin of the operation significantly increases the reliability and availability of your plant by continuously monitoring your mechanical assets using IIoT.

The realization of this needs cross-organization collaboration. Realization of the concept needs various tools, software, hardware and collaboration platforms to work together. This talk with explore those possibilities and readiness.

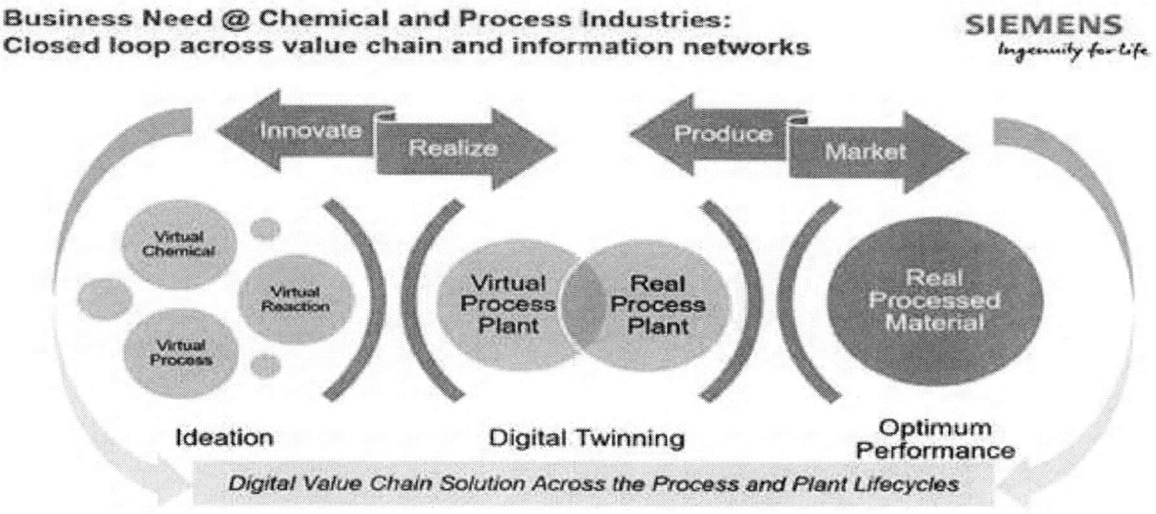

571319 Digital Twin for Aluminum Smelting

Building digital twins for manufacturing processes like aluminum smelting using first principles modeling has been considered challenging owing to the scale and complexity of such processes. A considerable mismatch is normally observed between measured data and predicted values of desired variables. This mismatch is usually because of uncertainty in the information fed to the model such as model assumptions, parameters, correlations from literature, etc. A systematic approach to identify such missing information would be to pose the model as an input-reconstruction problem. While a typical input reconstruction problem is solved using optimization-based algorithms where the objective would be to minimize the deviation between data and prediction, it may not be suitable for complex processes such as aluminum smelting. In this work we present a novel method to reconstruct missing information using a combination of first principles modeling and data analysis. The input reconstruction problem is reformulated as a set of differential algebraic equations (DAE) using measured data. Solving these DAEs help identify the missing input information and other unknown states in one shot. This method gives a complete data set (reconstructed inputs and measured data) which is used to train machine learning (ML) algorithms to build new correlations and predict unknown inputs directly in the digital twin model thereby giving more accurate predictions of desired process parameters.

558653 A Co-Teaching Approach to Unit Operations and Capstone Design - Joint Projects

In many programs, upperclass chemical engineering students take both their capstone design course as well as a course in unit operations. At the University of Connecticut, the Unit Operations (UO) course also doubles as the process simulation course - students learn about the theory of various unit operations as well as the basics of simulating them in software - in our case ASPEN Plus. The senior Capstone Design course is part of a two-semester sequence in which students work in groups on their chosen design project. The UO class and the first semester of design are offered concurrently in the fall semester.

Several challenges have arisen with this model in recent years. First off, the variety of senior design projects has blossomed. From faculty-sponsored projects evolving out of their research interests to externally sponsored projects from industrial partners, the projects that students work on are increasingly diverse, and do not necessarily lend themselves well to traditional process simulation. Secondly, guidance from ABET has suggested that departments need to be more diligent about assessing individual student achievement in design concepts. As most programs use a group-based approach to senior design projects, assessing individual achievement and competence in design topics can be challenging.

These two challenges - finding a common topic on which to build student knowledge in plant or process design and unit operations, as well as assessing individual achievement in design led us to develop an individually-based mini project that we used as a basis for instruction in both the Unit Operations course and Capstone Design. Students in design still had their year-long group project, but a key feature of the course was also an individual mini-project with multiple intermediate deliverables ***in both courses*** that developed knowledge in design, UO, and process simulation.

For our common mini-project we selected the Sabatier Process for *in-situ* resource production on Mars. The key reaction here is 4H2 + CO2 -> 2H2O + CH4. We divided the project into three main components: Upstream Processing, Reaction, and Downstream Processing. As an example of the structure of the project, Upstream Processing requires both storage of compressed hydrogen as well as compression of Mars ambient carbon dioxide. For the Design Portion of the course, this involved discussion of pressure vessel design as well as the design of the compression scheme. For UO/Process simulation, this involved simulating the compressors in ASPEN and selecting appropriate property models and thermal management. The students had related but different deliverables for each segment of the project due in both courses on the same deadlines, which encouraged them to think holistically about the project, but also allowed them to essentially apply the work from one project to two courses.

By collaborating on the Mini Project, we were able to expose students to multiple concepts across two courses and integrate the content in a meaningful way, as well as ensure that we were able to measure individual student achievement in design concepts across two classes. Each mini-project deliverable was clearly mapped back to both course and ABET objectives. We would encourage any program that has this type of structure to consider a joint project such as this.

577383 Undergraduate Chemical Engineering Laboratory: A Senior Capstone Experience

Successful capstone experiences engage senior students in an educational pursuit that ties together all components of the chemical engineering (ChemE) curriculum. The laboratory sequence at the University of Kansas (KU) serves as one such experience, giving students a deeper perspective on ChemE fundamentals and preparing them for careers in the chemical industry, research, or medicine. The first laboratory course focuses on the ChemE foundations of Transport, Kinetics, and Thermodynamics and uses three newly commissioned mobile experiment stations, specifically designed for use in the senior laboratory as well as for incorporation into their junior-level counterpart core courses. Integrating concepts from the laboratory into earlier courses provides for learning in multiple contexts over an extended period of time and helps students reinforce knowledge and develop understanding. The second laboratory course, taken in the final semester senior year, focuses on unit operations in process control and process separations. A new pilot-scale distillation column fabricated and donated by Koch Industries (Wichita, KS) allows students to study fundamental chemical engineering while also providing an opportunity to operate equipment more in line with that used in the process industry.

Response to continual advancements in the courses has been overwhelmingly positive, with students noting a greater understanding of chemical engineering concepts as well as better preparation for their future careers. This presentation will outline the recent improvements in the laboratory sequence at KU and its use in providing an immersive capstone experience.

577413 Wastewater to Beer: An IPA for the EPA

With the increased demand on water resources in arid regions and several notable droughts across the United States over the last decade, the efficient reuse of municipal wastewater has become increasingly important. Because beer making requires large volumes of water, several environmentally-conscious breweries worldwide have explored the possibility of using recycled wastewater as a feedstock. However, customers are often hesitant to consume recycled water products. To overcome the stigma associated with wastewater reuse, the Louisville Metropolitan Sewer District (MSD) has partnered with the University of Louisville to develop a water purification unit to supply local brewers who will produce small batch, special edition beers. As part of this partnership, seniors in the chemical engineering design course have designed and tested benchtop water purification units and performed calculations to scale their designs to meet MSD's needs. Students have explored topics in filtration, adsorption, flocculation, sterilization, and organic chemistry. The students' designs and recommendations have been incorporated into a final production unit that is being constructed at MSD's facilities. Five local breweries have agreed to participate, and their special edition beers will be served at a water treatment conference this summer. A class outline, student designs, water quality assays, and the results of a blind taste test will be presented.

572775 Chemical Engineering Lab for Seniors at United States Military Academy

The United States Military Academy has been ABET accredited since 2012. One of our capstone courses for senior chemical engineering students is CH459, Chemical Engineering Lab. This course has evolved since its inception to now include a total of six laboratory exercises, consisting of 6 lab periods each with an expected deliverable of a poster, executive summary, or full laboratory report. The rubric for each lab includes an independently developed lab experimental design/procedure with a minimum of three successful runs with usable data; chemical engineering process controls Laplace block diagram; data analysis and calculations to include mass and energy balance; Kline-McKlintock error propagation and error analysis; and a full CHEMCAD model of the process with plausible explanations of discrepancies. In addition there is a requirement for a face to face in progress review on the deliverable within 24 hours of completion of the last lab period, which allows for adequate coaching and mentoring of the scientific writing process.

The way in which the course is managed has allowed our chemical engineering seniors to develop acumen with respect to scientific writing and we have seen anecdotal improvement with respect to the quality of the poster, executive summary, and full lab report deliverables. Lastly, the students enrolled conduct peer evaluations on their lab team members to further their development as leaders and lab team members, and reinforce the concept of working as a member of a team.

556482 Applying PBL Methodologies to the Chemical Engineering Courses: Unit Operations and Modeling and Simulation, Using a Joint Course Project

Unit Operations and *Modelling and Simulation* have been for long time a staple of the academic formation of any chemical engineer. Both are integral to the analysis of any chemical process, discretizing it into smaller specific processes that can then be characterized and modelled through the solution of balance, thermodynamics and transport equations. However, students usually perceive these subjects as separate fields of knowledge, and they do not develop the ability to correlate and integrate them to solve real-world problems in their future profession.

On account of this, a Project-Based Learning (PBL) methodology was proposed in the redesign of the abovementioned undergraduate courses, focusing initially in Unit Operations. This PBL method was implemented alongside a joint course project, consisting on the design, assembly and characterization of a centrifugal pump, to be analyzed experimentally and computationally. To assess the success of this methodology, a survey was conducted on the students after they finished their courses. The results were mostly positive (85%), as the students appreciated the design component of the project, considering that it benefits their learning process, as well as the challenge it presented. This difficulty forced them to resort to different sources of information and areas of knowledge, alien to those provided in the courses. The limitations of the chosen project revolved around its limited scope and lack of connection with other topics of the courses (i.e. distillation columns). These limitations will be addressed with the design of transversal projects, which can cover more of the subjects seen in both classes.

577838 Developing Chemical Engineering and Analytical Science Acumen through Optimization of the Chocolate Process

Maintaining interest and motivation in science education continues to decline with increasing age [1]. However, extracurricular and guided inquiry laboratory experiences have been shown to promote interest [2]. Thus, the Chocolate Club, a sub-group of the United States Military Academy's American Institute of Chemical Engineers, combines chemical engineering and analytical science acumen and knowledge through the production of a consumable product, chocolate. Cadets design the chocolate process by identifying the product (such as 70% dark chocolate), ingredients, and specific process details such as mixing standards and refinement time. They extend their design to analyze products and compare between products using Fourier-transform infrared spectroscopy, thermogravimetric analysis, differential scanning calorimetry and rheology.

We will discuss the connection between the chocolate process and the chemical engineering topics of heat transfer, separations, batch reactors and kinetic analysis. We will also discuss the significance of the analytical topics of Fourier-transform infrared spectroscopy, thermogravimetric analysis, differential scanning calorimetry and rheology. This will be done with an example product of milk chocolate. The process will be discussed and connected with chemical engineering topics. We will then walk through the individual results of the analytical techniques and compare these results with various seed chocolates and refinement times (variations in the chocolate process). The analysis will focus on the comparison of the various procedures and the results of a blind taste test. The data analysis will also include fitting the rheological data to the Casson model using the genetic algorithm.

571993 Using Multidimensional Metrics to Assess Changes to Student Attitudes and Ability in a Capstone Laboratory Sequence

In many chemical engineering curricula, the capstone laboratory sequence is the primary source for students to get hands-on experience applying concepts learned in courses to actual pieces of equipment. The capstone laboratory provides essential instruction related to process safety, experimental design, data acquisition from modern instrumentation, analysis of results, and communication prior to a student's graduation and commencement of professional employment. However, capstone laboratory experiences can vary widely across institutions, and student perceptions of their learning experiences can vary widely from their actual knowledge and performance in the laboratory classes. As such, we have developed a mixed-method assessment of students as they progress through the laboratory sequence at Rose-Hulman Institute of Technology. The initial assessment was performed to gauge how students developed throughout a three quarter (one year) sequence in the laboratory to judge the strengths and weaknesses of our laboratory curriculum. The results of these assessments are intended to inform curricular revision at our institution and, hopefully, others as well.

This study was conducted over two years with two different cohorts of chemical engineering seniors. Students took a two-part assessment of their laboratory ability at the beginning of each quarter of the laboratory class and once at the end of Chemical Engineering Laboratory III. The first part of the assessment consisted of a version of the Undergraduate Research Student Self-Assessment (URSSA) that had been modified to be more relevant to teaching laboratory activities as well as the Music Model of Motivation (MUSIC). In the second assessment, students would take a laboratory skills test that would assess their understanding of instrumentation, flow diagrams, data analysis, and other laboratory-related skills. The results of these assessments, as well as a discussion of trends in student attitude and ability throughout the laboratory sequence, will be discussed.

579338 Overview of Chemical, Bioengineering, Environmental, and Transport Systems Division (CBET)

First of two sessions on initiatives at the National Science Foundation (NSF). This session will feature talks by NSF program managers on opportunities within their programs. Everyone is welcome to attend. NSF is an independent federal agency created by Congress in 1950 "to promote the progress of science; to advance the national health, prosperity, and welfare; to secure the national defense..." With an annual budget of about $8.1 billion (FY 2019), we are the funding source for approximately 24 percent of all federally supported basic research conducted by America's colleges and universities. In many fields such as mathematics, computer science and the social sciences, NSF is the major source of federal backing.

579339 Highlights of CBET Cluster on Chemical Process Systems

First of two sessions on initiatives at the National Science Foundation (NSF). This session will feature talks by NSF program managers on opportunities within their programs. Everyone is welcome to attend. NSF is an independent federal agency created by Congress in 1950 "to promote the progress of science; to advance the national health, prosperity, and welfare; to secure the national defense..." With an annual budget of about $8.1 billion (FY 2019), we are the funding source for approximately 24 percent of all federally supported basic research conducted by America's colleges and universities. In many fields such as mathematics, computer science and the social sciences, NSF is the major source of federal backing.

579340 Highlights of CBET Cluster on Engineering Biology and Health

First of two sessions on initiatives at the National Science Foundation (NSF). This session will feature talks by NSF program managers on opportunities within their programs. Everyone is welcome to attend. NSF is an independent federal agency created by Congress in 1950 "to promote the progress of science; to advance the national health, prosperity, and welfare; to secure the national defense..." With an annual budget of about $8.1 billion (FY 2019), we are the funding source for approximately 24 percent of all federally supported basic research conducted by America's colleges and universities. In many fields such as mathematics, computer science and the social sciences, NSF is the major source of federal backing.

579341 Highlights on CBET Cluster on Environmental Engineering & Sustainability

First of two sessions on initiatives at the National Science Foundation (NSF). This session will feature talks by NSF program managers on opportunities within their programs. Everyone is welcome to attend. NSF is an independent federal agency created by Congress in 1950 "to promote the progress of science; to advance the national health, prosperity, and welfare; to secure the national defense..." With an annual budget of about $8.1 billion (FY 2019), we are the funding source for approximately 24 percent of all federally supported basic research conducted by America's colleges and universities. In many fields such as mathematics, computer science and the social sciences, NSF is the major source of federal backing.

579342 Highlights of CBET Cluster on Transport Phenomena

First of two sessions on initiatives at the National Science Foundation (NSF). This session will feature talks by NSF program managers on opportunities within their programs. Everyone is welcome to attend. NSF is an independent federal agency created by Congress in 1950 "to promote the progress of science; to advance the national health, prosperity, and welfare; to secure the national defense..." With an annual budget of about $8.1 billion (FY 2019), we are the funding source for approximately 24 percent of all federally supported basic research conducted by America's colleges and universities. In many fields such as mathematics, computer science and the social sciences, NSF is the major source of federal backing.

579343 Interactive Question and Answer Session with NSF Program Directors

First of two sessions on initiatives at the National Science Foundation (NSF). This session will feature talks by NSF program managers on opportunities within their programs. Everyone is welcome to attend. NSF is an independent federal agency created by Congress in 1950 "to promote the progress of science; to advance the national health, prosperity, and welfare; to secure the national defense..." With an annual budget of about $8.1 billion (FY 2019), we are the funding source for approximately 24 percent of all federally supported basic research conducted by America's colleges and universities. In many fields such as mathematics, computer science and the social sciences, NSF is the major source of federal backing.

577532 Best Practices for Teaching Focused Faculty

Between 2006 and 2015 the number of full time teaching focused faculty in Engineering and Technology programs has increased by more than 80 %, yet during approximately the same period the number of full time Tenure/Tenure track faculty grew by about 10 %. Statistics will be provided for the number of chemical engineering faculty in the US by rank, position and gender. An overview of the contributions that can be made by Teaching focused faculty will also be presented. Based on the literature and experiences of Teaching focused faculty, best practices are recommended for the successful, mutually beneficial practices for Teaching focused engineering faculty and their administrative leadership.

580825 Cultivating Support and Success As a Department's First Permanent Non-Tenure-Track Faculty Member

In the fall of 2015, I was hired as the first permanent non-tenure-track faculty member in the Department of Chemical & Biomolecular Engineering at the University of Delaware. The position has brought many challenges but also many opportunities, and I was able to build on my past experiences as a visiting teaching assistant professor and non-tenure track lecturer at two other institutions to help define my position. Over the past four years, I have established roles in my department that make my job fulfilling and full of opportunities for growth.

I believe the University of Delaware is one of the "gold standards" in terms of logistics related to non-tenure-track faculty. Our treatment of this nontraditional track is evident even in the name: we are called Continuing Track (CT) faculty, to define us as something that we are (continuing from contract to often-lengthening contract), rather than something we are not (i.e., on the tenure track). CT professors share titles with tenured/tenure-track (T/TT) faculty, as well as a parallel promotion path; they are afforded sabbaticals every six years; they participate in faculty governance. Our status is further secured through the support of the American Association of University Professors, as well as a CT Caucus that works to improve the quality of life for all CT faculty at the university.

Being the first CT faculty member of a department came with its challenges: the right balance of teaching, scholarship, and service needed to be agreed upon, and the department faculty had to update our by-laws to align with the policies for CT faculty at the university and college level. I have relied on mentorship from CT faculty in adjacent departments (particularly chemistry and mechanical engineering) as well as T/TT faculty in my own department, as no one person has wholly worked under the same expectations as I have. My department chairs have worked carefully with me to ensure that I am given a workload that allows me to thrive as an educator. In this talk I will share elements of the creation and cultivation of my position as an assistant (and now associate) professor on the Continuing Track in chemical engineering at the University of Delaware that I believe to have impacted my ability to excel in ways that are different from, yet parallel to, our T/TT faculty.

558658 Lessons from an Untenured Administrator…

As the number of non-tenure track faculty members grows across the country in engineering disciplines, it is perhaps inevitable that some of us will end up in administrative roles that in the past may have been primarily reserved for those on the tenure track. Department head, assistant or associate department head, director of undergraduate studies, and even associate dean positions are now sometimes being filled by faculty colleagues that are members of a class of faculty known by many titles: instructional faculty, teaching faculty, in-residence faculty, teaching-track faculty, and the like. The names may be different, but the job descriptions are generally the same: we don't do "technical" research in our primary discipline (though we increasingly are getting involved in educational scholarship!), and we do a much larger share of teaching (and potentially service) than our counterparts on the tenure track.

In this presentation, I hope to present some of my observations (and best practices) that I've seen as a tenure track faculty member (2003-2008) as a non-tenure-track faculty member (2009-Present), and as an academic administrator: associate department head (2011-2013) and Associate Dean (2013-Present). NTT faculty can encounter a broad range of experiences in their departments based on both the micro-scale (departmental) and macro-scale (university-wide) culture, and this can lead to both some fantastic opportunities as well as feelings of isolation and overwork. Taking advantage of certain opportunities without the protection or stature that tenure grants can be daunting, but also incredibly liberating. It can also be frustrating when certain advancement paths seem closed to us because the position "requires" a tenured faculty member (for no reason that anyone can seem to clearly articulate). In academia, tenure is an obvious marker of quality and success, and one that has been collectively defined. For those of us not on that path, recognition and advancement can be more nebulous, and often more defined by our personal relationships than by anything else. I hope to share with you some of the ways that I have been successful in navigating the NTT world, and that some of those lessons are transferable to your circumstances and institutions.

580826 Presenting Yourself to the World: Academic Portfolio As a Professional Brand

Do you see yourself as a brand or just a busy academic focused on surviving your teaching, service, and research assignments? You may be working hard, but at the end of the day, a professional brand is the sum total of you and how you'd like to present yourself to the world – or at least specific target audiences (colleagues, students, employers, funders, promotion committees, etc.) The processes of sharing who you are can sometimes be overwhelming and exhausting. However, what if we create leverage our academic expertise and prepare a portfolio that can help us improve our academic career, showcase our work, and build a community in engineering education?

In this presentation you will be able to identify best practices on constructing an academic portfolio that can potentially become part of your professional brand. Also, your will learn how to share it online using multiple platforms. You are encouraged to bring a computer or any material that you want to share.

578186 Panel Discussion: Getting Directions on the Path Less Traveled

This segment of the session will be a panel discussion exploring the journey of non-tenure-track faculty from a variety of perspectives.

577031 National Academies Transforming Separation Science Report: Chemical Engineering Separations Curricula

An ad-hoc committee of the National Academies of Science, Engineering and Medicine (http://nas-sites.org/dels/studies/separations/) is concluding a consensus study on developing a research agenda for the next era of separation science. The committee included expertise from academia, National Laboratories and the private sector. As part of its statement of task, the committee was to "Identify the educational and human resource needs (including cross-sections of academia and industry) to enable advances in separations science." This presentation will report on the needs that were identified, as relevant to chemical engineering separations curricula. Included in the presentation will be results of supplemental survey on Separations Science that was developed and conducted by Professor David L. Silverstein and Professor Margot A. Vigeant for the AIChE Education Division, which helped to inform the National Academies committee. At the time of abstract submission the study was under review and will be published in May, 2019.

574878 Regional, Institutional, and Departmental Factors Associated with Gender Diversity Among Chemical and Electrical Engineering Graduates

Engineering remains the least gender diverse discipline within the science, technology, engineering and mathematics (STEM) fields. Chemical engineering (ChE) and electrical engineering (EE) are exemplars of relatively high and relatively low gender diversity, respectively. Here, we investigate the departmental, institutional, and regional factors associated with gender diversity among ChE and EE BS graduates at 95 institutions within the US, 2010 – 2016. For both ChE and EE, gender diversity was significantly higher at private institutions and at historically black colleges and universities (HBCUs), but did not show a significant association with the gender composition of tenure-track faculty. The number and type of other engineering majors, as well as variations in the EE departmental name, also showed a significant association with gender diversity. States with a highly gender-diverse pool of full-time ChE and EE workers tended to have a more gender-diverse pool of BS graduates, suggesting a contribution from regional variation in gender stereotypes. State financial support of K-12 education significantly impacted gender diversity among both ChE and EE BS graduates, with spending on "support services" have more of an impact than spending towards instructional services. State-specific gender pay disparity did not show a significant association with gender diversity among graduates. Nationwide, gender diversity among EE BS graduates appears to be increasing, while ChE diversity appears stagnant. Some regional variation in trends was also observed. This identification of factors associated with increased gender diversity may enable increased participation of all under-represented groups in a variety of STEM, and non-STEM fields.

577227 Implementing a Small-Scale Heat Transfer Radiative Experiment for Classroom Implementation

A heat transfer course offered in the third year of chemical engineering focuses on a foundational understanding of the three modes of heat transfer: convection, conduction, and radiation. Conduction and convection are often covered in great detail, with radiation linked to our few encounters with it in the real world: the sun's warming of the earth and the temperature of a fire. This method leaves gaps in a student's understanding of when and how radiation contributes and can be the dominating mode of heat transfer in a system. This study reports on transforming a previously conceptualized concentric annulus geometry designed to emphasize the effect of radiation. The inner tube is filled with sand and a thermocouple to monitor temperature and is wrapped with a resistance wire connected to a DC power source. The outer tube can be left clear or coated with a silvered tape, changing the emissivity of the surface. Student lab reports consist of two deliverables: first, questions intended to probe their understanding of system components and how to configure the unit to provide data for convection, conduction, and radiation, and second: collecting data with analysis of when the unit was at unsteady and steady state, and reporting with evidence the temperatures at which each mode of heat transfer dominated the system. Feedback data of the implementation is reported, with expected student comments to reflect a better intuition of when to recognize and check if radiation is important to a system.

571669 Knowledge Laboratories: Approach to Facilitate Independent Inquiry and Experiential Learning for 2nd Year Students

Knowledge Laboratories: Approach to facilitate independent inquiry and experiential learning for 2nd year students

Pavan Inguva[1], Mingrou Xie[1], Umang V. Shah[1]*, Clemens Brechtelsbauer[1]

[1*] Department of Chemical Engineering, Imperial College London, South Kensington Campus, London SW7 2AZ, United Kingdom

Abstract

Introduction

The learning paradigm of students progressively changes throughout their higher education journey as they mature and develop the skillset and thought processes needed as graduate engineers. To facilitate this growth, educators should provide the appropriate scaffolding both throughout the whole higher education journey and at each individual step. The nature of the scaffolding can be understood through Vygotsky's zone of proximal development. Students in the four year integrated master's course at Imperial College have a laboratory component in each of the first three years.

In the first year (Foundation Laboratory), students are introduced to the laboratory setting and are taught essential good lab practices. In the third year (Discovery Laboratory), students are given a chance to experience a realistic research setting where they work with an academic member of staff and can propose both a topic and methodology for research. The second year serves as a bridge between these two. This growth can be understood through progression along two axis: fixed problem → fixed problem-open-ended solution approach → open-ended problem and solution approach. Figure 1 schematically represents each laboratory activity from 1st to 3rd year which strives to gradually develop students' ability to tackle open-ended.

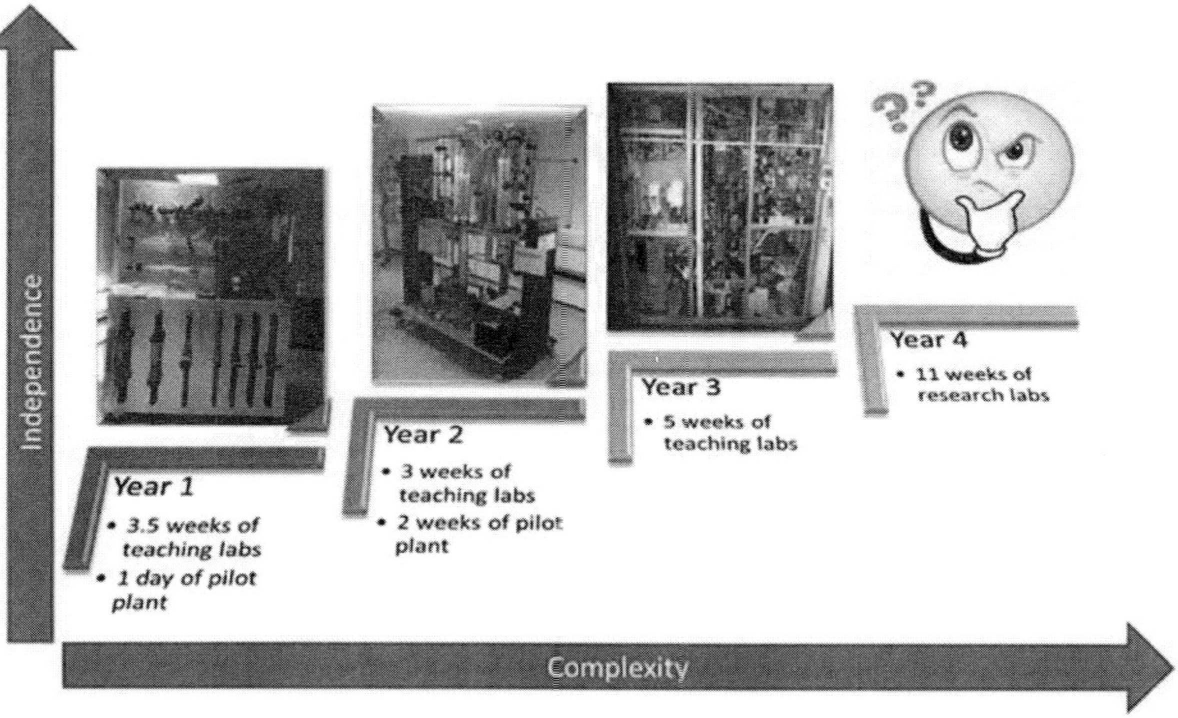

Figure 1 A journey into Discovery

In the past, Knowledge Laboratory took the form of a 'cookbook' recipe style course like the 1st year Foundation labs. Students had both a fixed problem and a fixed solution, but the problems provided and the level of analysis and rigor expected was commensurate with 2nd year students' capabilities. This approach was not popular with multiple cohorts with feedback indicating that students did not feel intellectually challenged and were not engaging with the course. Despite operational interventions from handout changes to the assessment strategy and GTA training exercises, student feedback did not improve. Upon comprehensive reflection of student feedback, the main issue was that the teaching philosophy of the Knowledge Lab was very similar to that of Foundation Lab. Correspondingly the Knowledge Lab experience was inconsiderate of the learner's zone of proximal development, and hence failed to engage students.

Here we present a case-study to illustrate the redevelopment process and pedagogical underpinnings of the new Knowledge Labs experience.

Course Context

This case-study was offered as one of 12 options that students could choose from all of which are structured similarly. Each module is broken down into three steps with step one, the introductory step, being at a difficulty level that is immediately accessible to the student and each step progressively becoming more challenging as they are exposed to increasingly newer and complex material. The course runs over three weeks, with each step taking one week.

In the case study, students are provided with a sample of degraded aspartame and are tasked to employ chromatographic techniques to purify and recover the aspartame. Step one tasks students to use TLC to analyze a sample of degraded aspartame that is to be purified. They could identify the degradation pathway and attempt to identify a suitable solvent composition for good separation. In step two, students need to develop a suitable method on an analytical HPLC system to quantitatively analyze the sample. They also need to explore various process conditions such as composition, mobile-phase flowrate and column temperature. In the last step, students utilize a preparative HPLC system to perform the separation at a larger scale and similarly optimize the system.

Pedagogic underpinning of module development

The pedagogical philosophy underpinning the restructure can be understood through a constructivist lens which has found broad application in scientific and engineering education, particularly in the lab (Abdulwahed & Nagy, 2009). Vygotsky (Vygotsky, 1978) characterizes the learner as someone who actively constructs their knowledge, with this construction taking place at the learner's 'zone of proximal development' by interactions with fellow students and teaching staff. This interaction can be activated through the provision of a well-structured course which includes for instance adequate communication and meetings between learners and teachers (Chen, Shah & Brechtelsbauer, 2016) and providing a sequence of planned activities that suitably guide students with the teacher acting as a facilitator whilst simultaneously providing students the independence and autonomy to pursue their own approach (Hodson, 1996).

The zone of proximal development is shown in Figure 2. The entire module can be thought of as an iterative activity that expands the domain of knowledge and activities that students can confidently do unaided. Each step therefore has been structured to simultaneously build up student's capability in the objective for that step and set the context for the next step. As each step is appropriately placed in the learner's zone of proximal development and each step helps to nurture students such that the next step is now within their zone of proximal development. For instance, using analytical HPLC before the module starts may be in the 'out of reach' zone, but by the end of step 1, HPLC techniques now firmly lies in the student's zone of proximal development.

Support and Scaffolding Provided to Students

The teaching philosophy of the module represents a significant shift away from the high school or 1st year 'cookbook' procedural style of labs to a more open ended and self-directed course structure.

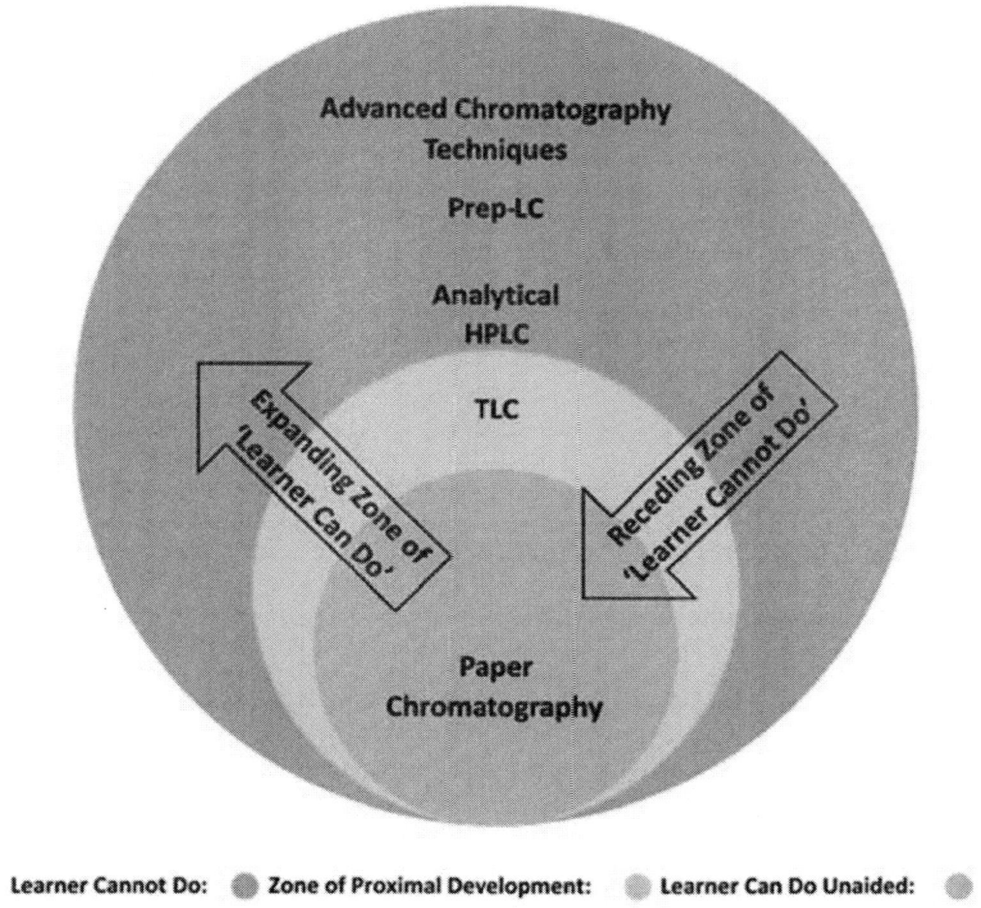

Figure 2. Student's Zone of Proximal Development at the Beginning of Step 1

This may be intimidating to students and as such, appropriate scaffolding and support needs to be provided along the way. Examples of the support given to students include:

1) Handouts: The handouts provided contain a problem statement and possible sequence of experiments to conduct for each step. These handouts contain references to relevant literature and are kept deliberately brief to nudge students to review literature independently.

2) Training videos on the operational aspects of the course such as setting up and running the TLC and the HPLC systems. Students are encouraged to view these videos before the start of the lab session so that they would have adequate operational knowledge and would be able to make full use of their laboratory time instead of focusing on operational issues.

3) Frequent review sessions with their project supervisor, a graduate teaching assistant (GTA). These provide an opportunity to review the work already carried out and to plan the next step. Students are strongly encouraged to pursue ideas beyond the provided suggestions from the handout. These ideas, together with the proposed workflow for the week are agreed upon with the GTA.

Evaluation of Effectiveness

The overall satisfaction of the module has been evaluated using the university's internal student feedback mechanism known as SOLE. Compared to 2015, the overall satisfaction towards the module has been observed to increase by ~35%.

A survey to understand student's response to the chromatography module (case-study) was carried out in 2019. Out of the 11 students who participated in this module, n=10 responded to this survey and the results are presented below.

The survey results prove that each module step was planned in an appropriate manner with students agreeing that the structure was logical and each step was well situated in their zone of proximal development. This is also reflected in student's perception of their skill level and their confidence to apply the techniques learnt in the future with almost all students indicating they are confident in using TLC and HPLC in their future practice. Lastly, students also mostly indicated that the level of support and scaffolding provided during the course was adequate for their learning.

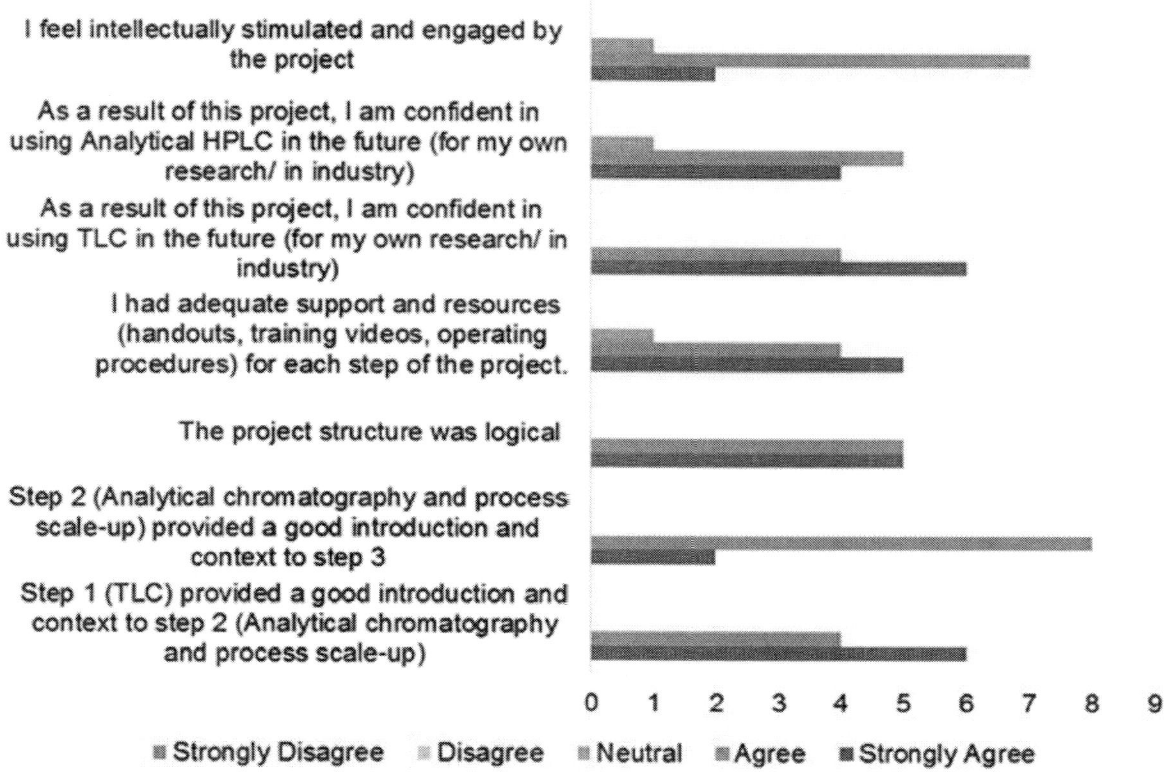

Figure 3. Chromatography Module Survey Results

Conclusion

This study evidently demonstrates an approach to facilitate independent inquiry and experiential learning in a laboratory based teaching projects. By empowering students with increased ownership of their projects and flexibility to propose and solve real-life process questions in an authentic learning environment, students feel intellectually stimulated and engage better with the course. This study highlights the important of appropriate scaffolding to ensure that students do not feel overwhelmed with such a jump from traditional 'cookbook' approaches to open-ended projects. Course design aspects like structure and teaching material also needs careful consideration in context of Vygotsky's zone of proximal development.

References

Abdulwahed, M. & Nagy, Z.K. (2009) Applying Kolb's Experiential Learning Cycle for Laboratory

Education. *Journal of Engineering Education*. 98 (3), 283–294.

Chen, W., Shah, U. & Brechtelsbauer, C. (2016) The discovery laboratory – A student-centred experiential learning practical: Part I – Overview. *Education for Chemical Engineers*. 17 (0), 44–53.

Hodson, D. (1996) Laboratory work as scientific method: Three decades of confusion and distortion. *Journal of Curriculum Studies*. 28 (2), 115–135.

Vygotsky, L.S. (1978) Interaction between learning and development. *Mind and Society*. pp.79–91.

564296 Improved Homework Effectiveness in a Problem-Solving Course

The effectiveness of homework in a junior-level, problem-solving course, specifically a kinetics and reaction engineering course, was assessed as the course format, homework grading basis and homework feedback mechanism were modified. Development of problem-solving mastery is central in the learning outcomes for this course. The learning literature establishes that (a) to develop mastery, students must acquire component knowledge and skills, practice integrating them and know when to apply what they have learned, and (b) goal-directed practice coupled with targeted feedback is critical to developing problem-solving mastery.

In this study, the distribution of total exams scores for students who completed 90% or more of the assigned homework (high completion group) compared to that for students who completed less than 90% of the homework (lower completion group) was used as a measure of homework effectiveness. Initially, the course was taught using a traditional lecture format with weekly homework assignments being graded and contributing to the students' course grades. Feedback consisted of returning "graded" assignments, posting a correct solution and providing office hours where students could seek assistance on pending homework and feedback on completed homework. Under these conditions, homework was completely ineffective; the difference in total exams scores between the high completion group and the lower completion group was not statistically significant. Results like these are not uncommon and have lead to questions on the usefulness of homework and to the development of automated homework systems.

The course was then changed to a flipped format; videos corresponding to the former in-class lectures were provided to the students for pre-class use. Much of the class meeting time was then used to engage students in learning activities that afforded them opportunities to practice problem solving in class with immediate instructor feedback and prior to being required to submit homework. Homework was still graded and contributed to the students' final course grades, and additional feedback again consisted of returning "graded" assignments, posting a correct solution and providing office hours. After flipping the course, homework appeared to be significantly more effective: the total exams scores of the high completion group were approximately one letter grade equivalent higher than the scores for the lower completion group. The precise reason for the increase in homework effectiveness is not known.

While the flipped format afforded additional problem-solving practice, the instructor suspected that students approached homework with the wrong goal. Specifically, their goal was to obtain the correct answer. The fact that their homework was graded on correctness and counted toward their overall course grade justified and reinforced their focus on obtaining a correct answer. Indeed, a focus on obtaining the correct answer on homework is deeply ingrained after many years of having homework graded and counted toward final course scores. By mimicking worked examples, students can efficiently obtain a correct solution, but in doing so, they often fail to learn when to apply and how to integrate their component knowledge on their own. For homework to be most effective, the students' focus and goal must be learning how to obtain the answer, not just obtaining it.

On the basis of the preceding reasoning, the course was again modified so that homework assignments were graded on the basis of apparent effort, and not on being correct. At the same time, each homework assignment was followed-up with the required submission of a self-corrected solution. The corrections were assigned immediately after the homework was originally due and the correct solution had been posted. Effectively, this approach attempted to have the students self-generate feedback on their homework. They still were provided with a correct solution, but their homework was not graded. Instead, they were expected to determine where and why their solution diverged from the correct solution.

This change did not improve the effectiveness of homework as had been intended. The difference between the high completion group and the lower completion group remained nearly the same. The instructor believed that the students' focus was still improperly directed. When they corrected the homework they submitted, they used the posted solution to determine where they went wrong, but they didn't reflect upon why they had gone wrong. Instead, they simple replicated the posted solution from that point on to complete their required corrections.

Most recently four changes to the course were implemented in an attempt to appropriately shift the students' focus as they completed homework assignments. First, for each of the six major type of problems considered in the course, explicit, in-class instruction on the identification of the problem type was introduced. Second, at the same time, a generalized procedure for solving each type of problem was explicitly presented together with the instruction on problem-type identification. Third, scaffolding that emphasized the generalized procedure for solving each problem type was incorporated into related in-class, problem-solving activities. Finally, instead of requiring students to submit a corrected homework solution, students were required to submit a homework wrapper. The homework wrappers ask the students to write a few sentences assessing the problem-solving strategy that they conceptualized in comparison to the generalized procedure that was discussed and practiced in class. After these four changes were implemented, the effectiveness of homework again increased substantially. The total exams scores of the high completion group were approximately one and one-half letter grade equivalents higher than the scores for the lower completion group.

The results of this study suggest that conventional homework can be very useful in helping students develop problem-solving mastery. They suggest that homework, at least initially for a given problem type, should be viewed and implemented as an opportunity for goal-directed practice and targeted feedback, and not as a summative assessment instrument. The results further suggest that when homework wrappers are coupled with explicit instruction on generalized problem solution strategies, they may enable students to self-generate constructive and targeted feedback on their problem-solving efforts. An additional benefit of this approach is that amount of grading support needed can be decreased substantially when homework is graded on effort.

560097 Work in Progress: An Undergraduate "Introduction to Research Methods" Course for Honors Students

In an effort to improve student recruiting and retention, our institution has focused on expanding the Honor's College experience. Within the College of Engineering and Applied Science, the goal is to focus the Honors experience on undergraduate research. Both Honor's Colleges and undergraduate research are considered High Impact Programs (HIPs) for recruiting and retention. The aim of the Engineering College Honor's program is to broaden research opportunities and competitiveness of student applications for summer research programs, NSF REUs, internal/external research funding applications, participation in undergraduate research conferences, and preparing the students for graduate school. Historically, many students have received credit for completing undergraduate research, but this is often a "stand-alone" course with no additional preparation and ill-defined outcomes. The new College Honor's program is also expected to lead to a significant increase in the number of students performing undergraduate research. In an effort to improve the undergraduate research experience and achieve the above outcomes, the College will be initiating an Introduction to Research Methods course for the undergraduate Honors students as pre-requisite for undergraduate research during the Fall 2019 semester.

The goal of this course is to group mentor the students on many common research topics in advance of them entering the laboratory and performing research. This should reduce the required mentoring load of the research advisor and also lead to above desired outcomes. The course will be broadly focused by providing a general approach to research and graduate school preparation. Course topics will include: finding a research mentor, literature search skills, using the scientific method for approaching a research problem, developing a research methodology, research ethics, laboratory safety, writing a funding proposal, delivering a research presentation, and selecting and applying for graduate school. This presentation will discuss the motivation for this work, course details, learning objectives, course schedule, and course assignments. In addition planned assessments and student outcome tracking for the course will be presented.

572185 Participatory Design As a Tool to Develop Advanced Experiential Learning Projects for Chemical Engineering Students

Participatory Design as a tool to develop advanced experiential learning projects for Chemical Engineering students

Pavan Inguva, Assen Batchvarov, Umang V. Shah*, Clemens Brechtelsbauer

*Department of Chemical Engineering, Imperial College London, South Kensington Campus, London SW7 2AZ, United Kingdom

Abstract

Introduction

Empowering students to become co-creators of their educational experience is one of the desired goals of student-centric teaching approaches. They traditionally focus on content delivery and course operations with student involvement being limited to providing feedback at the end of the course. We have previously used Participatory Design (PD) as a tool to design experiential learning modules.

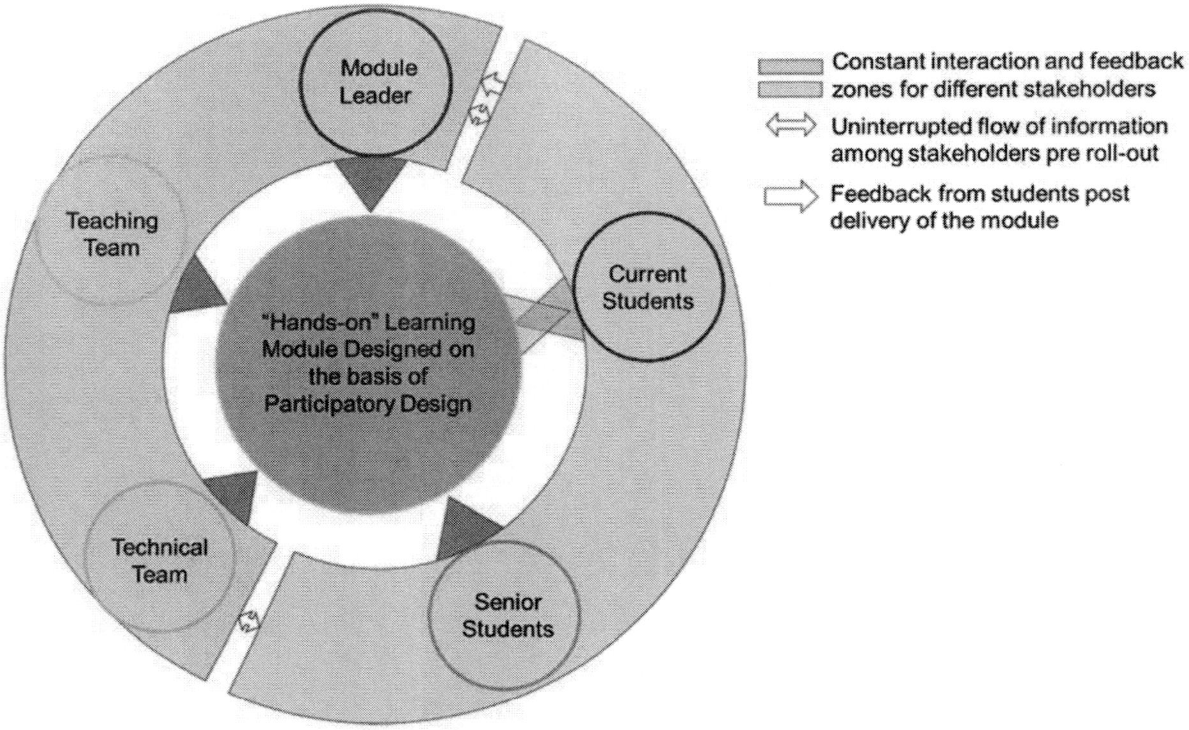

Figure 1 Schematics explaining participatory design based approaches for experiential learning based module development.

PD proved to be effective in enhancing student engagement and module teachability. By involving undergraduate (UG) students of different degrees of seniority and teaching assistants (TA) as partners in the module design process, their experiences and feedback can be integrated from the conception stage right up to delivery.

The use of participatory design for developing modules with fixed learning objectives is evident in the literature (Inguva *et al.*, 2018), however, employing similar approaches for open-ended projects poses its own unique challenges. Research based (semi-)open ended projects require advanced teaching tools appropriate to the learning level and a module structure capable of enabling open-ended enquiry. This study aims to develop a PD based approach for semi-open ended, research based modules designed with experiential learning as the pedagogic underpinning.

Research Method

This investigation part of ongoing efforts of intervention based pedagogic research, commonly known in literature as action research. It involves students as key stake holders in shaping their and their peers' learning. The Key pedagogic research question asked for this work is "How does a participatory design (PD) based approach to design an open-ended research project change from a PD based approach for projects with fixed objectives?"

We used the participatory design approach for module development to develop and implement a third-year core chemical engineering laboratory project themed around mixing in stirred vessels. The project was designed to allow students to independently propose and investigate a variety of influencing factors.

The project was developed over the summer of 2018 involving two second-year UG students, two third-year UG students, a graduate teaching assistant (TA), a teaching lab technician, and an academic with relevant expertise in the area.

The effectiveness of the PD based approach was evaluated through a weekly questionnaire to track development of the stakeholders. Focus group interviews and a survey of all the students, who were offered this project were also conducted at the end of the module.

Evaluation of Effectiveness

Due to the small number of students who were offered this module (n=10), the survey focused on collecting student feedback in the form of free text comments.

Students broadly acknowledged and appreciated the open-endedness and flexibility afforded by the set-up and the teaching tools. They also enjoyed the ability to further adapt the tools provided to suit their needs.

"[A] *Good number of options were available, helping to understand different aspects of mixing*"

"[The] *Set-up was easy to understand and modify compared to [sic] other experiments*"

When asked about their perspectives on involving students in the module design process, the students who took the module felt that this brings the following benefits: 1) Contemporary design and innovation, and 2) enhanced accessibility.

To holistically evaluate the value of PD, on-top of learner experience, the dividends to stakeholders (skills development for student partners and development of educational resources for educators) need to be considered.

Dividends to Stakeholders

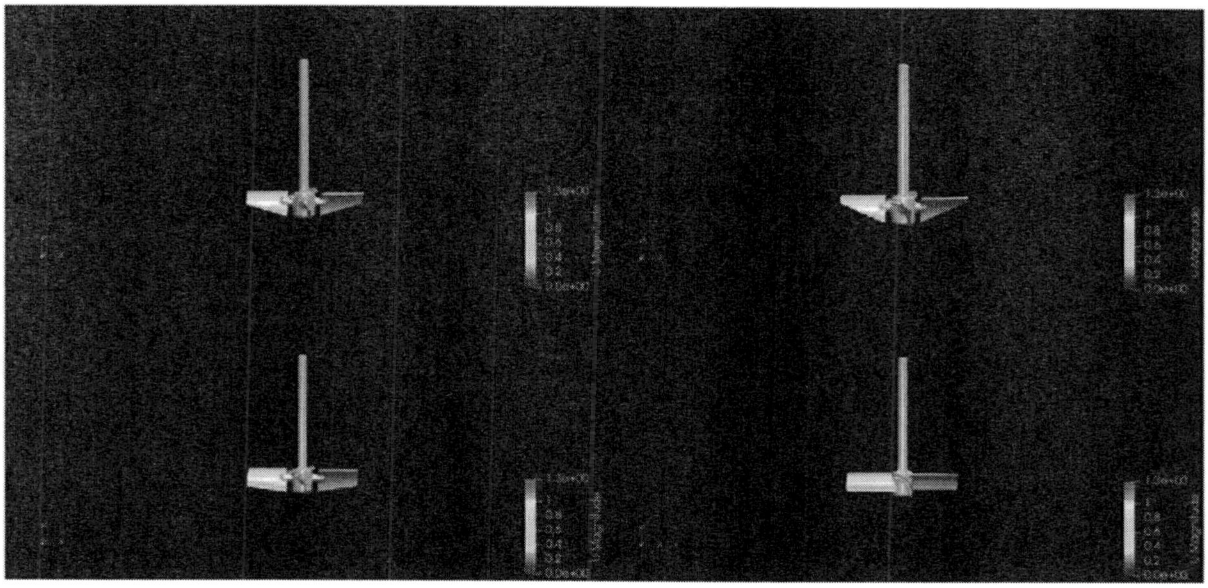

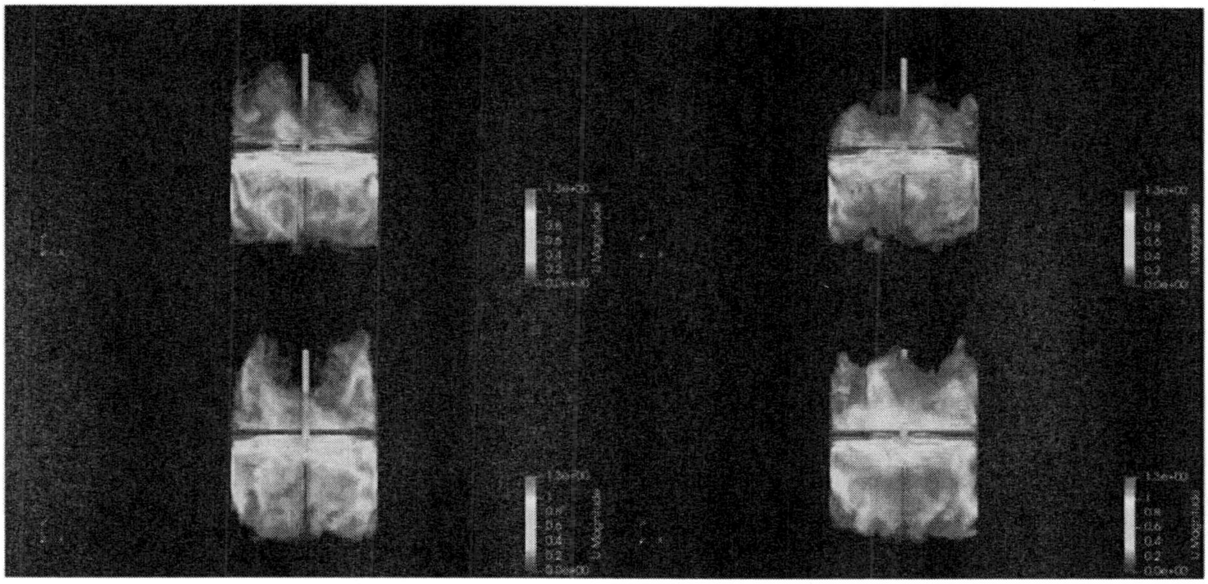

Figure 2 CFD model with different impeller geometries (Top) and sample simulation results (Bottom)

To provide prospective students with multiple analytical tools for exploring the design space, the 3rd year students and TA proposed implementing a Computational Fluid Dynamic (CFD) model of the system. In addition, one of the 2nd year students proposed an image analysis tool for studying mixing and implemented it using MATLAB. The development of these sophisticated tools requires expertise and skills in specialist areas. Whilst most teaching staff may have a good theoretical understanding of these topics, they may not have the skill or time to implement such tools.

Student stakeholders involved in the module development process were expected to learn and employ newly acquired knowledge to design experimental resources. It was observed that students were independently learning new skills, technical knowledge and could creatively apply the acquired knowledge with limited guidance. Any module designed with a PD framework should have a training component. For example, the development of the CFD system was carried out in openFOAM, a highly powerful open-source CFD software. By working closely with the experienced TA, the students gained proficiency in this tool. Examples of the model developed by the student partners are shown in Figure.2.

The construction and commissioning of the set-up can also provide valuable learning opportunities to student partners. Figure 3 shows the CAD file done by one of the 2nd year student partners who was keen on learning CAD and machining. The student took significant initiative in liaising with suppliers and the in-house machine shop, gaining proficiency in industrially relevant engineering skills. The high quality of educational resources developed demonstrate the capability of student partners to grow and contribute given the right support and training.

This emphasises a broader project management point that educators should take note of when implementing a PD framework. Stakeholders need to be engaged at a suitable skill level and be given the help they need to grow. Ostensibly, as the skill level required for the task increases (from junior to senior level module development), educators should be aware of this in selecting students and providing training resources.

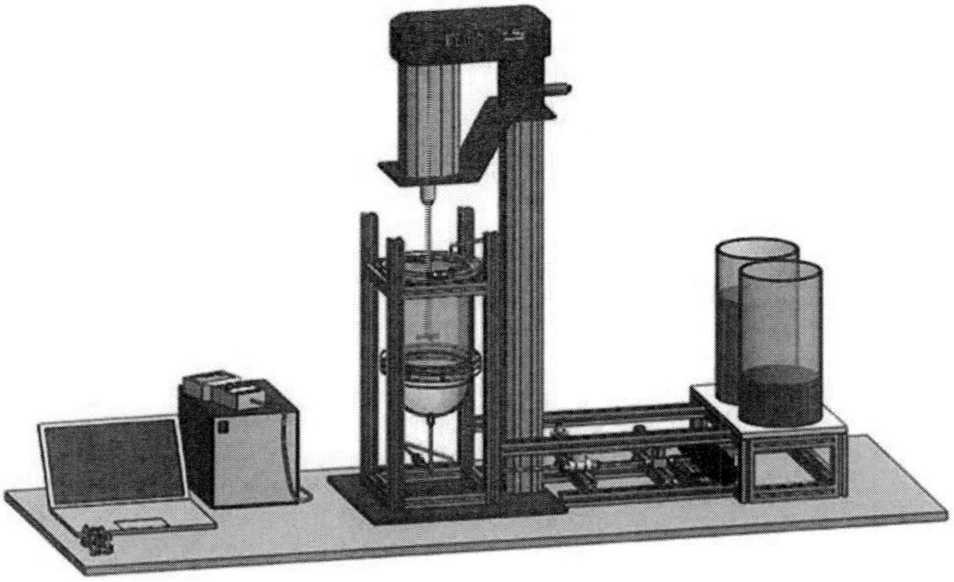

Figure 3 Computer-Aided Design (CAD) Drawing of Set-up Made by Student

Participatory Design – Changing Roles for Stakeholders

Discussion with participating stake holders suggested that the broad ideas of a PD based module development approach still are applicable to the module design process for open ended research based module. On a macro-level, each stage of the module design process (from conception to construction and beta-testing to roll-out), benefits from 1) the stakeholders' perspective of how the module and teaching material can be improved for both educators and students 2) providing unique ideas and suggestions based on their personal or professional experience. On a micro level, incorporating a double feedback mechanism where suggestions from the educator and other stakeholders can be iterated upon, ensures that ideas are exhaustively evaluated. One key point was that free flow of information between different stakeholders was fundamental to the success of the process. Students felt that they could express their views in a more relaxing and open environment, where discussions were focused on intellectual merit and benefitted from the collated expertise of the module development team.

Subtle changes in stakeholder interaction and involvement was observed compared to the participatory design based module design approach reported in the literature. In addition to the specific example cited above, the following changes in the interaction dynamics between stakeholders were observed:

The educator motivated the student stakeholders to take control of the project, providing input on broader pedagogical and process issues rather than being involved in development minutiae. Rather than directly intervening and solving problems, the educator helped students to find the resources or contacts they needed. In summary, the educator assumes the role of a facilitator, providing an enabling environment to the student and TA stakeholders.

As the project development progressed, the student stakeholders felt empowered to conceptualize aspects of the module themselves. Considering the open-ended inquiry desired by the module, the student stakeholders now provided the process, ideas, and the specifications for tools to implement them. These would be discussed and agreed on by all the stakeholders including educators to ensure appropriate quality.

Conclusions

Our work validates PD as a module design framework to develop highly student centric modules, including open-ended research. Educators need to be aware of the increased technical complexity of these modules and how this correspondingly influences stakeholder dynamics for effective project management and stakeholder engagement.

References

Inguva, P., Lee-Lane, D., Teck, A., Anabaraonye, B., et al. (2018) Advancing experiential learning through participatory design. *Education for Chemical Engineers*. 25, 16–21.

579344 Proposal Writing Tutorial

Second of two sessions on initiatives at the National Science Foundation (NSF). This session will feature talks by NSF program managers on opportunities within their programs. Everyone is welcome to attend. NSF is an independent federal agency created by Congress in 1950 "to promote the progress of science; to advance the national health, prosperity, and welfare; to secure the national defense..." With an annual budget of about $8.1 billion (FY 2019), we are the funding source for approximately 24 percent of all federally supported basic research conducted by America's colleges and universities. In many fields such as mathematics, computer science and the social sciences, NSF is the major source of federal backing.

579345 Interactive Breakout Panels

Second of two sessions on initiatives at the National Science Foundation (NSF). This session will feature talks by NSF program managers on opportunities within their programs. Everyone is welcome to attend. NSF is an independent federal agency created by Congress in 1950 "to promote the progress of science; to advance the national health, prosperity, and welfare; to secure the national defense..." With an annual budget of about $8.1 billion (FY 2019), we are the funding source for approximately 24 percent of all federally supported basic research conducted by America's colleges and universities. In many fields such as mathematics, computer science and the social sciences, NSF is the major source of federal backing.

558649 Teaching Aspen Plus Via Student-Developed Video Modules

Teaching process simulation tools such as ASPEN Plus can be a challenge in some departments. If there are not faculty that regularly use the software as part of their research, then the learning curve for the novice faculty can be steep. Additionally, process simulation software is often used in the senior or capstone design course, and teaching it concurrently often does not allow students enough time to develop proficiency with many of the features and capabilities the software has. Spreading the instruction out over multiple classes so students develop proficiency over time is an appealing idea, but this then requires faculty in multiple courses to develop expertise in the software and/or give up time in their class for process simulation instruction.

To address some of these challenges, we have designed a senior-level elective course with the goal of having students develop ASPEN-focused educational training videos and other educational materials for deployment in a modular form throughout the chemical engineering curriculum. In our pilot test of this class, students who had successfully completed the department's Unit Operations and Process Simulations class and showed superior proficiency with ASPEN were recruited to participate. Students in the class were required to develop at least two instructional videos between five and 10 minutes in length that covered specific topics in ASPEN. As part of the elective course, students were introduced to topics in engineering education and assessment. Students were encouraged to select topics from the chemical engineering curriculum that corresponded to topics they had seen in their previous coursework and were of personal interest to them. Working with the faculty, the students developed a clear set of learning objectives for their modules, and then developed the content of the videos with those learning objectives in mind. Students developed a "shot list" and script for their videos, and created a draft video that was then critiqued by the faculty and their peers. Students then created a final version of their video incorporating this feedback. In addition to the video, students were encouraged to develop in-line questions or concept quizzes that can be embedded in the videos to check student comprehension and promote active engagement with the video. Lastly, students were required to develop a problem set with solutions that were reflective of the learning objectives that they outlined at the beginning of their video.

The end result is a series of modules with clearly stated learning objectives and solved problem sets that faculty in multiple courses can "drop-in" to their classes without sacrificing significant in-class time to software instruction or developing significant expertise in the software itself. Over time, this creates a library of modules for use in various courses throughout the curriculum, allowing students to engage with process simulation earlier, and providing an always-available resource for them to return to in the future. This also allows new videos to be made over time as topics change or the software gets updated. The elective class itself encourages students to develop their lifelong learning skills and professional presentation skills, as well as engage in significant amounts of peer learning.

560239 Integration of Aspen Plus® Videos into the Chemical Engineering Curriculum at the University of Kansas

The University of Kansas (KU) Department of Chemical and Petroleum Engineering (C&PE) is integrating Aspen Plus® into its core chemical engineering courses from Freshman to Senior year. The aim of this integration is to supplement the teaching of traditional course material and homework problems, while providing a marketable skill for chemical engineering students post-graduation. This presentation will highlight current methods of introducing Aspen Plus® into the curriculum, which include online videos, written handouts, lecture notes, and homework assignments. The presentation will demonstrate how the videos were created using Camtasia Studio 9® screen capture software, as well as the advantages of using this innovative tool for instruction and enhancing the learning experience. Student surveys will be discussed, which have assessed the effectiveness, interest, and performance of using Aspen Plus® in the classroom. Ultimately, the goal of the speaker (Junior C&PE undergraduate) is to fully implement Aspen Plus® into the 4-year chemical engineering curriculum at KU for current and future students, in addition to creating an online library of videos available to all students taking chemical engineering courses.

559231 Teaching Chemical Engineering Fundamentals with Comsol Multiphysics®

COMSOL Multiphysics® provides simulation and analysis of processes described by partial differential equations solved with finite element methods. It is used extensively in both industry and academia for fundamental and applied research in chemical engineering as well as many other disciplines. It is also widely used in teaching mathematical modeling courses at many universities. Although not yet widespread, it has recently been introduced in undergraduate courses on transport phenomena and reaction engineering to help teach fundamental chemical engineering principles. In this presentation we describe our experience using it for teaching in a unit operations laboratory course and a sophomore level chemical engineering thermodynamics course.

In our UO laboratory course we use simulations as pre-lab exercises to review fundamentals and prepare for the lab. Students can also use the simulations to help analyze and augment their experimental results. Fluid flow simulations illustrate differences between compressible and incompressible flow, as well as laminar and turbulent flow, and show that friction losses through pipes and fittings measured in the lab can be calculated from fundamental principles. A heat transfer simulation not only correctly predicts outlet temperatures of a heat exchanger, but also shows the velocity and temperature profiles throughout. Other interesting simulations include a PID temperature control process for a jacketed chemical reactor, carbon dioxide removal from an air stream in an absorption process, heat transfer in beverage bottles, and hazardous heat effects of exothermic reactions.

For these pre-lab exercises students normally follow a detailed tutorial to work directly with the software to make changes in the model parameters and rerun a pre-built model at various prescribed conditions to gain insight into the process studied. Recently, we introduced some pre-lab exercises that were delivered as "apps" over the internet via COMSOL Server™. The apps allow students to adjust model parameters and observe numerical and graphical results via a control panel for a virtual experiment without going into the COMSOL Multiphysics software. The results of a comparative study of the "tutorial" method and the "app" method with regard to student satisfaction and learning will be presented.

COMSOL introduced a thermodynamics feature with version 5.3a at the end of 2017. This feature allows calculation of thermodynamic properties and phase behavior for non-ideal chemical mixtures. We developed apps for generating and interpreting VLE phase diagrams, understanding the physical significance of activity coefficients, analyzing a pressure swing distillation process, and studying the factors that affect chemical reaction equilibrium and implemented them in our mixture thermodynamics course. Most apps were assigned as homework and contained quizzes that could be automatically graded and submitted via email directly from the apps from any internet browser. The chemical reaction equilibrium app was implemented as a virtual experiment that students used to complete a team project to design a recycle reactor/separator process. Our experience with the initial implementation of these apps will be discussed.

578321 Smart Computing in Undergraduate Courses - Improved Understanding of Fundamentals

Experience in using a user-friendly software, Mathcad, for problem solving in undergraduate courses to improve understanding of process fundamentals is discussed. The problem solving involves the following steps: (a) pose the problem, (b) formulate the problem, (c) analyze the problem, as much as possible, without assigning numbers using symbolic algebra techniques, (d) decide what numerical approach will be the most efficient to solve the problem, and (e) only then proceed with numerical solution. Students learn interactively by changing parameter values in the problem to observe variation in problem solution. Students are encouraged to display results using graphics for improved learning and understanding. Emphasis is on obtaining solutions with increased confidence and higher precision and with minimum effort by understanding what type of problem is being presented. Students are encouraged to not introduce unnecessary complications when attempting to solve a problem and solve problems using optimal effort. Simple problems are solved using simpler techniques. More complex techniques are left only for solving more complex problems. Students learn faster with simulations and trying, failing, and succeeding on their own than via lectures, reading textbooks, and being evaluated via exams and quizzes. Mathcad has user-friendly syntax allowing the user to display the problem as if doing written work in a notepad using pencil/pen. Results are displayed on the go as parts of the problem are typed. Error flags allow for quick corrections of mistakes as and when they occur. Mathcad is interactive, with user-friendly graphics capabilities. Students can change numbers and see variations in results for the problem they are solving. In this manner, students can gain further insight into process characteristics by trying problem solving on their own. Pertinent example problems are considered for illustration.

576362 Integrated Molecular Modeling Education in the Chemical Engineering Curriculum

Given the critical importance of thermodynamic and process modeling, these concepts have been successfully integrated into the chemical engineering curriculum. However, as in-silico design gains credibility in engineering, molecular modeling is becoming more useful in chemical engineering. Molecular modeling with never replace experimental synthesis and characterization, just as process modeling will never completely replace pilot plant experiments. However, molecular modeling can provide insight into the design of optimal chemical and material systems, just as process modeling provides a basic design before final optimization of the process. It is important for chemical engineering students to see the value of molecular modeling in their field as its use continues to grow. Unfortunately, the integration of molecular modeling into the chemical engineering curriculum lags behind that of process simulation despite its increasing importance. We have studied molecular modeling education in various classes with chemical engineers including: a specialty elective in applied molecular modeling, molecular modeling modules in a chemical product design course and a chemical engineering numerical methods course, and molecular modeling modules in a polymer science course and laboratory. Our experience has taught us that the most effective approach to teaching this subject employs active learning in which the students carry out applied molecular modeling exercises in class and for assigned projects. The biggest challenge using this approach is a lack of scaffolded computational infrastructure. Process simulation software like ASPEN, or numerical simulation platforms like MATLAB or COMSOL possess an ease of entry for students due the their advanced user interfaces. Chemical engineers find it more challenging to use academic molecular modeling programs that have more primitive interfaces. This is analogous to the command line interface of the original academic version of ASPEN. Some commercial platforms such as Biovia's Materials Studio have efficient easy-to-learn user interfaces, but do not allow sufficient access to the underlying algorithms and models to facilitate fundamental learning by the student.

We have found that the Molecular Operating Environment (MOE) from Chemical Computing Systems addresses both these challenges. MOE allows students access to fundamentals while simultaneously providing a high level vector-based coding language to write their own simulation and analysis applications without having to code low-level geometric and energetic routines. These qualities allow us to carryout active learning in the classroom where students can run, code, and analyze molecular simulations in class to learn both fundamentals and relevant engineering applications. Also, because MOE integrates across the relevant scales of molecular modeling from quantum mechanics, through molecular modeling, and up to meso-scale models, it can be used to show students the importance of multi-scale modeling in this field. We will present examples of how this integrated approach works in practice using our current classroom model that employs both active learning and a flipped classroom mode. Our flipped classroom displaces the typical lecture on modeling fundamentals from the classroom to pre-class videos. This allows us to carry out actual molecular simulations and analysis in class to illustrate how molecular modeling works and why it is valuable. Without this optimal platform such an approach would not be possible. Further, we utilize some of our recent work on cognitive load theory to formulate optimal modeling examples in this class. Previously, we showed that the optimal choice of a programming examples in a numerical methods course can maximize the germane cognitive load (mental effort used in learning), while minimizing extraneous cognitive load (mental effort that distracts from learning). We illustrate how this optimal choice can potentially improve learning outcomes in a numerical modeling classes.

564859 Combining Various Software Tools for Enhancing Programming and Numerical Problem Solving Capabilities

This paper deals with the calculation of the Adiabatic Flame Temperature (AFT) for the combustion of gas mixtures in air with a variety of software packages and data sources. This will provide insight into how different mathematical software packages can be used as students proceed through their Chemical Engineering studies. The overall problem statement, needed data, useful equations and several solutions are given in an internet link.

At the Ben-Gurion University, first year students are required to take the introductory "Modelling and Computation" course where the mathematical software packages: POLYMATH (http://www.polymath-software.com) and MATLAB (http://www.mathworks.com) the Excel spreadsheet and the DIPPR (https://dippr.aiche.org) physical property database that is typically provided with POLYMATH are taught. The "Modeling and Computation Course" is described in detail by Shacham (2005). In a more advanced "Process Simulation" course, MATLAB and DIPPR are used for modeling individual unit operations. UniSim (https://www.honeywellprocess.com) process simulation program is used for simulation of complete processes. This course is described in more detail by Shacham (2011).

The use of the different software packages will be demonstrated here by calculating the "Adiabatic Flame Temperature (AFT)" for Ethane and Methane natural gas mixtures (presented as problem 2.13 in Cutlip and Shacham, 2007, a copy of the problem definition is available at the ftp site: ftp://ftp.bgu.ac.il/shacham/AIChE_19/Appendix_A.pdf) . Physical property data (heats of combustion and heat capacity equations and values) are typically available from classical thermodynamic books. The solution of most problems usually start using POLYMATH whose desirable properties are mentioned, for example, by Mahecha-Botero et al. (2011) who advise that the program is appropriate "for solving systems of algebraic and ordinary differential equations because it requires minimal computer programming skills and is extremely simple to use."

Exercise 1 (1st year CHEG students)

In this exercise the students are required to formulate the POLYMATH program for calculating AFT for one particular value of x (th air-to-fuel ratio) and a value of y (inlet ethane concentration). The POLYMATH program, which is self-explanatory, is shown in Figure B-1. (Note: the Figures and Tables won't be included here in this paper, but they are available at the FTP site ftp://ftp.bgu.ac.il/shacham/AIChE_19/Appendix_B.pdf). The solution for AFT involves solution of a nonlinear algebraic equation (NLE) which can be easily solved by pressing the pink arrow when the problem coding from Figure B–1 has been entered. Solutions obtained are, for example, AFT = 2198.0 K, for air-to fuel ratio of x=0.5 and 2503.3 K, for $x = 1$.

Exercise 2 (1styear CHEG students)

The next exercise is to obtain and plot AFT values for various air to fuel ratios and natural gas compositions. MATLAB and Excel are more suitable for such an assignment, consequently the POLYMATH options of exporting the problem statement to MATLAB and Excel are used.

A typical MATLAB function (as generated by POLYMATH) for this Exercise is shown in Figure B-2. The students are required to prepare a main program that calls the MATLAB function and carries out the calculations for large sets of x and y values. The requested MATLAB program is shown in Figure B-3 and the resulting plot of AFT vs air-to- fuel ratio, generated by this program, is shown in Figure B-4. Students find that learning MATLAB programming by modifying an existing, working program is much easier than starting from scratch.

In solving Exercise 2 with Excel, an error-prone and difficult task is to replace the POLYMATH variable names in the data base equations by appropriate cell addresses. This can be expedited by exporting the

POLYMATH code to Excel as shown in Figure B-5. In column "B" of the Excel spreadsheet the original variable names are shown and in column "D" the original equations. The modified equations, in which the variable names were replaced by their addresses, are inserted in column "C". The solution for AFT is obtained by using the "Solver" or "Goal Seek" tools to minimize the contents of cell "C19" by changing the value that appears in cell C18. Starting with a correct spreadsheet representation of the problem definition, it is necessary to generate tabular and graphical representations of the AFT results for different parameter values.

Exercise 3 (3rd year CHEG students)

The accuracy of the physical property data (heat of combustion and heat capacity correlations) included in the problem statement can be compared with the AFT obtained using the "Conversion Reactor" unit of the UniSim commercial process simulation program. The results of the UniSim calculations are shown in Figure B-6, for the parameter values x =. 1 and y = 0.75. The exit temperature from the conversion reactor is 2044 °C = 2317 K. The AFT temperature calculated, using the POLYMATH program for the same conditions is: 2503 K. Thus there is 8.03 % difference between the two solutions.

It is assumed that the significant difference between the AFT calculated by UniSim and the Excel model is caused by the difference in the precision of the physical property data used. This assumption can be checked by replacing the heat of combustion data and the heat capacity correlations in the POLYMATH model are replaced with up-to-date data from the DIPPR Physical Property Database. Note that a subset of 113 components from the DIPPR database are accessible from POLYMATH. This subset includes the six compounds included in the exercises.

The POLYMATH program, in which the heat of combustion data and the heat capacity correlations were replaced with the data and correlations from the DIPPR database, is shown in Figure B-7. The reported uncertainty for the heat of combustion data of CH_4 and C_2H_6 is: < 0.2 %. The uncertainty for the ideal gas heat capacity correlations for all the compounds is reported as < 1 %. The AFT calculated, using the program shown in Figure B-7 for the parameter values $x = 0.\ 1$ and $y = 0.75$, was AFT = 2313 K, thus the difference from the value obtained by UniSim is only 1.12%. Thus, using more accurate physical property data brings the solution much closer to the value obtained with UniSim.

Conclusions

The problem solving of the Adiabatic Flame Temperature (AFT) for Ethane and Methane Mixtures in Air provides a real system for a host of calculations that can be conducted with a variety of commonly used PC software packages and process simulators. Use of POLYMATH, Excel, MATLAB, DIPPR Database and the UniSim Process Simulator are integrated into a progression of problem solving of increasing complexity that can be introduced progressively within the Chemical Engineering curriculum. Students enjoy this relatively gradual approach that builds up their capabilities and understanding of both the problem and the required problem solver that is being used. The POLYMATH export to MATLAB and Excel considerably speeds up the learning of these software packages. The good results of the comparison with a Process Simulator's results (with its internal data base) increases the student's| confidence in the correctness of the problem solutions and helps to verify the results.

References

1. Cutlip, M. B. and Shacham, M. *Problem Solving In Chemical and Biochemical Engineering with Polymath, Excel and MATLAB.* Prentice-Hall, Upper Saddle River, New-Jersey, 2007.
2. Mahecha-Botero, A., Reaume, S., Grace, J.R., Ellis, N., "Independent research as a teaching tool in graduate chemical reaction engineering. Case study: modelling isomerization of unsaturated fatty acids with catalyst deactivation". Educ. Chem. Eng. 6 (1), e1–e9 (2011).
3. Shacham, M., "An Introductory Course of Modeling and Computation for Chemical Engineers". Comput. Appl. Eng. Educ. 13, 137-145(2005)
4. Shacham, M., "Use of Advanced Educational Technologies in a Process Simulation Course".

Computer Aided Chemical Engineering, 29, 1135-1139(2011).

565051 A Scaffolded Approach to Process Modeling Education with Literate Programming Tools

Mathematical models allow the designer to optimize and predict, and so process modeling is a core within any chemical engineering program, either as a separate course or embedded across the curriculum. Process modeling education has students use mathematical and computational skills to develop, test, and interpret models. To succeed, students must rebuild and refine their knowledge from introductory courses in programming and numerical methods plus also use material from their chemical engineering core: heat transfer, diffusion, chemical kinetics, etc. The reality is, of course, that past coursework can rapidly disappear into the mists of time and subsequent instructors must help the students become reacquainted with it.

A common approach to the teaching of engineering modeling is to focus on the use of numerical methods techniques with a small amount of simple programming skills being needed to deliver the required results. Understanding the principles of the model, developing skepticism based on the "garbage in, garbage out" principle, and robustly interpreting the results are key steps at this early stage that then permit successful engagement with sophisticated simulation packages in later stage courses.

For teaching purposes, the mathematics and its embodiment in code are insufficient on their own, with inclusion of the context of the problem, a description of the approach in words and interpretation of the results being necessary. Jupyter notebooks offer a literate programming environment in which numerical methods and data analysis code can be interactively developed and refined in 'cells', while relevant rich context including text, images and equations is interspersed amongst the code. For the engineer undertaking one-off model development and also in the educational scenario, the Jupyter notebook offers the ideal fusion of code, documentation and output.

To scaffold the three-fold development in programming, numerical methods and discipline knowledge, we have created Jupyter notebooks in which problem statements (in Markdown, with images included and mathematics via MathJax) are accompanied by code cells (in Python or Octave). As well as the problem statement, questions come with some hints on possible approaches or fragments of code as starting points to help students. A small amount of guidance, without giving away the entire solution, allows students to attempt much harder questions and develop their skills. Within each week's tutorial set, progressively less help is offered as confidence and competence is gained.

To facilitate maintenance of the tutorial notebooks, the problem and solution notebook are generated out of the same annotated precursor/template notebook using some simple Python modules we have developed; the notebook with the problems is generated by redacting parts of the code to leave some partial solutions or fragments as hints, omitting the output of the code, and omitting the analysis of the model output. After editing the instructor-facing template to move questions, fixing bugs, improve explanations etc, the student-facing problem and solution sheets are both automatically regenerated.

Beyond the educational advantages of this scaffolded approach in the classroom setting, we have also found that marking take-home assignments where students work in outline templates was faster than where students work completely independently outside of the template. While little assistance is offered in these assessment activities, the notebook imposes some additional structure onto the work and may also hint towards using certain variable names, reducing the time required to decide on the grades for each piece of work, thereby allowing the time to be spent offering better feedback to the students. Jupyter notebooks can be exported to PDF and these are marked within our learning management system's PDF grading tools; alternatively, tools like nbgrader can work with the notebooks directly.

Hosting and deployment of any computational tool presents interesting challenges, particularly for large class sizes. The benefits and limitations of cloud-hosted Jupyter notebooks at CoCalc.com will be discussed and the use of the templating system will be demonstrated.

559282 Introducing Chemical Engineers to Coding, Data and Visualization through Matlab

This talk will describe our use of Matlab in teaching chemical engineering undergraduates the principles of coding in an elective course as well as in independent research projects. Our approach focuses on the use of Matlab to solve traditional chemical engineering problems as well as its use to general interest science, technology and economic topics with an emphasis on simulations, data analysis and visualization. A number of student projects will be highlighted showing the creative ways in which students have applied the coding skills they have learned in as little as one semester, several of which have contributed to peer-reviewed journal publications.

577933 Teamwork and Teamwork Evaluation Skills

Engineering curriculum is interspersed with opportunities for students to engage in teamwork. But often students are placed on teams with no instruction on how to be a successful member of a team. In addition, faculty aren't often trained on how to observe or evaluate teams. As a result, teamwork is often graded on the final deliverable of the team and not the teamwork skills that students should be developing.

In this talk the presenter will share favorite literature and resources on teamwork and teamwork evaluation skills. Examples of teamwork skills collected from the literature include establishing a team charter or team expectations, using team roles, and countering unproductive behaviors. Tips on observing and evaluating teams will also be shared.

562425 Using the Trolley Problem in Chemical Engineering Ethics Education

The so-called trolley problem has been used as a common tool of ethicists to introduce difficult topics of morality to the classroom and to the public. In this problem, and its many variants, participants are asked to participate in a choice that will either result in the hypothetical death of or injury to one person versus several people. In many instances, respondents have been found to gravitate to choosing harm to the individual over the group. However, depending on how the questions is posed and how the people involved are described, the majority of respondents may switch to favoring harm to the group. Using functional magnetic resonance imaging studies, researchers have found the decision between harm to the individual verses the group engage distinct regions of the human brain, and may reveal important moral decision-making processes.

These same processes are often in play in the daily life of a working chemical engineer, when attempting to assess the hazards and risks associated with their decisions in the design process or during daily operations. As automation increases, engineers are finding themselves in the position to code these ethical choices into the scripts needed to make decisions when a human operator cannot make them in real-time. As such, implications from these automated choices need to be considered well in advance by all engineers, from those designing the next autonomous vehicles, to those designing the next generation of oil refineries.

For ten years, we have used the trolley problem (and its variants) within the engineering ethics portions of our senior Unit Operations Laboratory to introduce concepts of "personal" verses "impersonal" ethical dilemmas. In this talk, we will discuss the results from the use of this common tool of ethicists, and how it can be used in a chemical engineering classroom settings to help students realize the influences on their ethical reasoning that often go unnoticed, even when they can dramatically alter the ultimate consequences to coworkers or the public.

578314 Preparing Students for Professional Success: Cross-Disciplinary Lessons

An important objective for many disciplines at academic institutions is to impart students with the specific skills and viewpoints needed to succeed in their chosen profession. Each discipline has carefully considered how their courses can best prepare students for professional life and has come up with strategies for ensuring their students practice the necessary skills. While in some cases the techniques used by a particular discipline are unique to that discipline, it is likely that there are commonalities between the approaches. It may be possible for educators in chemical engineering, for example, to learn from faculty in journalism on how to prepare students for the professional world. In this project, a survey was sent to faculty in 46 different departments at Kansas State University to evaluate how they taught students the skills they would need in their professional life. In addition, 33 different classes were visited, representing 23 different academic departments. This presentation will describe both quantitative and qualitative results on the ways in which faculty impart the skills their students need to succeed as professionals. Both commonalities and differences will be described, with specific attention to unique ways of training students that may be transferable between disciplines.

562135 A Novel Course Sequence on Critical Thinking, Publishing, and Publication Ethics for Engineering Graduate Students

We teach two, one-credit graduate courses for engineering (STEM) graduate students at the University of South Carolina aimed at improving the ability of graduate students to publish peer-reviewed papers and to do so while conforming to ethical standards of publication and research. The first course, which is for new graduate students, teaches critical reading and analysis of the research literature in their field. In this first course we also address the ethical issues of plagiarism and how to properly cite, paraphrase, and summarize information. The second course, for graduate students preparing to publish, teaches students how to write a journal paper or proposal for review. The ethical issues addressed in this second course are authorship, review standards and processes, and avoiding data falsification and data fabrication. The theoretical underpinning of both the reading and writing courses is the Paul-Elder model of critical thinking (Paul & Elder 2012).

We will present the critical thinking model along with the course materials including course syllabi, typical assignments, resources for publication ethics, and methods for effectively assessing student work. We will specifically discuss how the courses have been designed to work within the constraints of typical graduate degree program requirements and within the resource constraints of engineering graduate programs.

Paul R, Elder L. 2012. *Critical thinking : Tools for taking charge of your learning and your life*. Boston: Pearson. 493 pp.

AGILE MANAGEMENT OF PHD PROJECTS: A RISK-MITIGATION TOOL FOR GRADUATE EDUCATION

Laura Pirro, Ghent University, Ghent, Belgium
Marina Montero Carrero, Ristretto.black, Brussels, Belgium
Francesco Contino, Université Catholique de Louvain, Louvain-la-Neuve, Belgium
Joris W. Thybaut, Ghent University, Ghent, Belgium

Introduction

Pursuing a PhD can feel like walking on a tightrope: exhilaration and adrenaline arising from the challenge can drive one forward, but the fear of failure and of a ruinous fall might be paralyzing [1].

The psychological distress faced by graduate students has multiple roots, such as difficulties in maintaining a work-family balance, lack of inspirational guidance, uncertainties about future career outcomes, etc. [2]. Addressing, not to mention solving, these challenges cannot be done with a one-size-fits-all strategy; each PhD is a unique experience by a unique individual [3].

However, one common feature is shared by all research projects, namely the strive to push boundaries of science forward. This is a highly uncertain and sometimes unpredictable task and for this reason traditional management strategies, such as 'waterfall' methods [4] and Gantt charts [5], tend to fail. PhD projects do not follow the plans and so far there is no clear risk-mitigation system to support the graduate student when this happens.

An alternative, *Agile* philosophy for research management was already advocated a decade ago [6], but, being focused on the challenges and the structure of the software industry, it failed to hit the big time in Academia. More recently the implementation of an *Agile* methodological workflow is finally raising interest in Academic environments [7],[8]. However, the variety in activities and challenges to be tackled during a long-term research project, such as the ones carried out by PhD candidates, still requires a comprehensive assessment of the advantages and potential pitfalls of *Agile*.

In the present work an *Agile* approach focused on graduate education, in general, and the management of PhD projects, more specifically, is proposed. The goal is to draw the attention of a broad Academic audience to this management philosophy, as a promising instrument for potentially enhancing the productivity of research projects, while at the same time improving the psychosocial wellbeing of graduate candidates.

Agile workflow for research projects

When the *Agile* Manifesto [9] was first published, 18 years ago, it started a disruptive revolution in the world of software development. One of the core principles at the heart of this philosophy is 'Responding to change over following a plan'. Having an *Agile* mindset means being flexible, responsive, adaptable. In an *Agile* world, progress is measured through the attainment of concrete results and working outcomes, rather than by ticking off successive tasks on a to-do list. An early, limited result, which can be swiftly elaborated in a later stage, is more significant than a perfect result achieved only by the deadline of the project [10]: the first allows for feedback during the project, while the latter simply does not. It is tempting to transpose the characteristics of *Agile* management as described above to graduate research projects.

Based on the available literature [11], some guidelines could be proposed to translate the concepts of *Agile* into practical guidelines for the daily activities of PhD researchers [12]. The suggested steps to be implemented are the following:

1. A big block of work is divided into several 'layers' of activities. What defines a layer is that it ends with the attainment of a concrete result, as opposed to the traditional waterfall methodology where activities are organized purely based on the timeline. Each layer will be undertaken in a specific, limited period of time (e.g. 2-12 weeks), named a *sprint*.
2. *Sprint* planning: the PhD student, the supervisor and any other stakeholders (for example a postdoc, industrial partner, or masters student) meet for around 30 minutes aiming to set the goal of the sprint and its duration. Everybody must agree on these two points, to ensure that the entire research team is on the same page and that prospects are the same for everyone. As part of the sprint planning, the sprint review meeting (see step 5) can already be scheduled.
3. *Sprint* execution: work! Maximum focus is dedicated on each particular task for a definite amount of time.
4. Weekly '*scrum*': meeting between the PhD student and the supervisor for a maximum of 15 minutes, but ideally every week to keep the momentum, (e.g. the same timeslot every week, outside of 'conventional' working hours to make sure there are no meetings, teaching activities etc. which may get in the way). The point of this meeting is to be short and efficient: better avoiding chairs so that everyone stands up and there is no risk of getting too comfortable and overextending the meeting. No laptops, not even papers. Only three questions need to be answered:
    a. What was done the previous week in order to contribute to the goal?
    b. What will be done next week in order to contribute to the goal?
    c. Are there any impediments?
5. *Sprint review + sprint retrospective + sprint planning*: at the end of the sprint, a meeting is planned including the PhD student and all the stakeholders involved to discuss the obtained outcomes and whether those are in line with the goals and expectations set up in the previous planning (review). This is the moment to go into the specifics and brainstorm altogether. It is crucial to discuss the encountered difficulties, in order to make sure the next sprint is continuously improved compared to the previous one (retrospective). This is the phase when impediment removal (= problem solving, in *Agile* jargon) is dealt with: sincerity and transparency are key. As previously mentioned, *Agile* is all about adapting to change, and plans can change. Go back to point 1 and restart the planning, addressing the next layer of work in a new sprint.

The conceptual difference between a traditional and an *Agile* approach to PhD management is summarized in Figure 1.

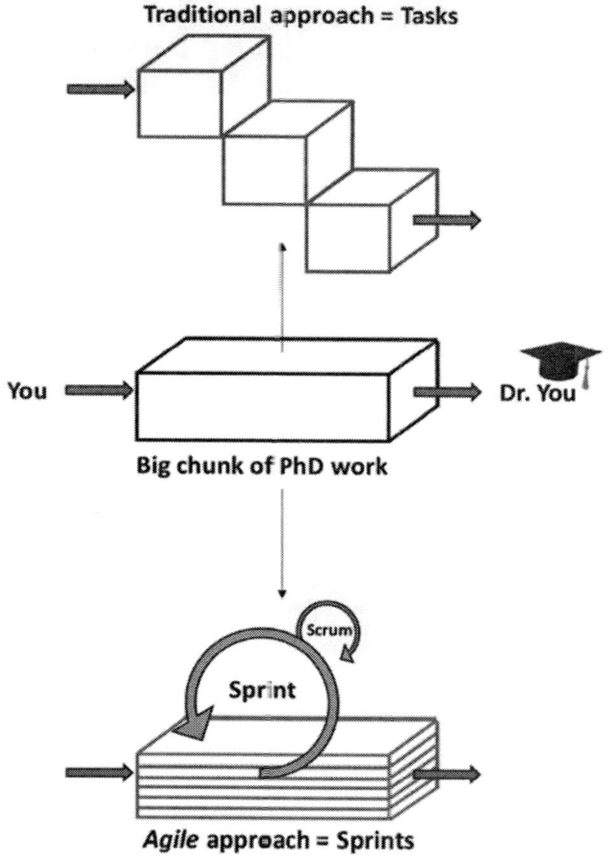

Figure 1. Conceptual difference between traditional PhD management (red) and *Agile* PhD management (green).

Academic *Agile* case studies

To illustrate the proposed approach, we provide a sample overview of the application of the *Agile* principles to three typical tasks of a researcher (for example, in chemical engineering). The simplified case studies reported in Table 1 offer a conceptual view of some activities which can be encountered in a research project and are meant to exemplify how the *Agile* approach aims at finalized, even if preliminary, results in the early stages of work.

Table 1. Comparison of traditional and *Agile* approach for three recurring case studies in graduate education of, among others, chemical engineers.

Long-term goal	**Traditional approach**			***Agile* approach**		
	Scope of task 1	**Scope of task 2**	**Scope of task 3**	**Goal of sprint 1**	**Goal of sprint 2**	**Goal of sprint 3**
Writing a manuscript	Write 'Methodology' and 'Results'	Write 'Discussion' and 'Conclusions'	Write 'Introduction'	Write down the key concepts of each section as bullet points; sketch figures of the 'expected' results	Add actual figures and tables with captions; include main references in each section	Elaborate text around the bullet points and the figures
Building a simulation code	Write detailed subroutines for input/output	Write detailed subroutines for calculations	Test the code	Write and test simplified input/output/ calculation subroutines	Increase the level of detail of the main subroutines and get them to work independently	Integrate all the subroutines in one working code; testing
Performing an experimental campaign	Design of Experiments (DoE) over the full range of variables and operating conditions	Execution of all the experiments	Data processing	Simplified DoE, execution of selected experiments and data processing	Increase the number of factors in the DoE, execution of new experiments, data merging and processing	Increase the number of levels per factor in the DoE, execution of new experiments, data merging and processing

In order to move towards realistic, applied case studies, we recruited researchers, students and professors willing to apply the proposed *Agile* workflow to a short-term project [13]. By remotely helping the participants to set-up the case studies and following up the progress vis Skype updates, several collaborations with other Academic institutions were established. Three examples are described below.

Writing a research proposal - University of North Carolina at Chapel Hill (May – August '19)

Who: Bailey DeBarmore is a senior PhD Student at the department of Epidemiology of the Gillings School of Global Public Health, under the supervision of Prof. Wayne Rosamond, who is also responsible for educational activities in the same department.

What: Bailey had approximately 14 weeks to draft and submit a full research proposal to obtain funding for her dissertation project. She had to start from the literature review, then move into developing and refining the specific aims, all the way down to a finalized text.

Why *Agile*: Driven by the desire of reaching her goal in the most efficient way, Bailey recognized that the *Agile* principles could respond to her need of a structured working method, while at the same time retaining a certain flexibility, as required by innovative research projects.

How: Bailey split her work in 4 sprints, which were guided by Wayne in weekly scrum meetings of approx. 30 min. For each of her sprints, she set a finalized written document as goal to achieve, continuously building up on previously written text and transitioning, as of the first moment, from a draft to a finalized proposal.

Feedback: -) No negative points were highlighted by Bailey and Wayne.
+) The case study was considered successful, with the proposal being handed in timely before the deadline. Wayne could appreciate a considerable difference compared to previous students he had coached in the past. In fact, as opposed to 'just' brainstorming for long time, Bailey's *Agile* approach gave him the possibility to always discuss and provide feedback over a concrete document.
Both Bailey and Wayne confirmed that they are likely to implement *Agile* in their future work.

Summer project with bachelors students – University of California, San Diego (June – August '19)

Who: Bill Connacher is a PhD student in materials science and engineering at the Jacobs School of Engineering – UCSD, in prof. James Friend's lab. Cesar Collins and Aneth Martinez worked under Bill's supervision at UCSD as part of the ENLACE 2019 summer research program for high school and college students.

What: Cesar and Aneth had to design and manufacture a capillary wicking structure within a Surface Acoustic Wave (SAW) device and had approx. 7 weeks to accomplish their goal.

Why *Agile*: Bill had acted as mentor for undergraduate students in the past and had encountered some difficulties in facilitating their achievement of a finalized result. In order to avoid the pitfall of uncompleted outcomes at the end of the internship, Bill decided to adopt an *Agile* workflow in mentoring Cesar and Aneth.

How: 3 sprints, which were managed via daily scrum meetings. The complexity of the work in each of the sprints was gradually increased, starting from a simplified laser ablation system, which, due to originally unforeseen project requirements, was later on upgraded with an additional capability.

Feedback: -) Bill encountered considerable difficulties in keeping the scrum meetings short and only once per day. This was attributed to the total lack of previous knowledge of the students about the research topic and it highlights one of the major requirements/drawbacks of *Agile* implementation: the need of a team which is highly skilled in the field. It can be expected that, in some cases, this might hinder adoption of *Agile* by starting PhD students and or early-stage researchers.
+) As opposed to previous internships, Cesar and Aneth's work resulted in a finalized structure which, despite not being optimized, was capable of effectively transporting fluid. Bill appreciated the opportunities arising from *Agile*, such as the incremental approach in defining goals and the flexibility in adjusting expectations, and he will try to implement a comparable workflow in some aspects on his own research.

Computational Fluid Dynamics simulations – CORIA, University of Rouen (June – September '19)

Who: Julien Reveillon is professor of Computational Fluid Dynamics (CFD) at the University of Rouen Normandy and researcher at UMR 6614-CORIA in the field of multiphase flows. Cléante Langrée is a PhD student, funded by INERIS, working under Julien's supervision. Javier Anez is a postdoctoral researcher affiliated to the CNRS, collaborating in Julien's group. Nathan Mauger is a graduate student at ESITech Rouen and he carried out a summer internship in CORIA with guidance from Javier and Julien.

What: Cléante and Julien: comparison of premixed flames CFD solvers, using flame velocity, flame thickness and adiabatic temperature as outputs of interest from 3D Large Eddy Simulations;
Javier and Nathan: numerical simulations of a boat's hull movement with various water aeration rates, with a specific focus on boat's drag and stability issues.

Why *Agile*: Julien's commitment to innovation in teaching and learning sparked his interest towards *Agile*. His concern is that too little time and efforts are dedicated to the development of professional skills and working ethics in engineering graduate programs. He saw *Agile* as an opportunity to challenge himself and his co-workers with an alternative, and potentially promising, approach to research.

How: In both cases, 5 sprints with a duration of 2 weeks each, were foreseen and managed via 1-2 scrum meetings per week. Each of the sprints had three goals: solve at least one of the several numerical challenges, write/update a written report on the achieved progress, summarize the sprint results into a structured presentation.

Feedback: -) Estimation of the time which would be actually required for a given task was often arduous. As a consequence, this could lead to demotivation and struggle. This is one of the most problematic features of research work, since the issues which are being tackled in research projects are usually new and/or not fully understood, and it represents an onerous challenge for any management style. Nathan found that he could partially mitigate this difficulty with the *Agile* workflow, by re-negotiating the goal of successive sprints in the review and planning phase.
+) In both cases an improvement in the mentoring relationships was observed, both from a professional and a personal point of view. Frequent interactions, even if shorter, resulted in enhanced communication and overall higher efficiency in the work. Enforcing the writing/presentation since from the first moment, each sprint was always concluded with a finalized work which served as solid starting point for the next one. Julien and Javier will continue their journey into Agile with their future projects. Having experienced the benefits in his own work, Nathan, who will supervise an undergraduate project during his upcoming final year as grad student, will try and adopt an *Agile* coaching style.

Conclusions

The application of an *Agile* management style to research projects could positively impact the Academic environment in terms of increased flexibility and adaptation to the ever-changing challenges to be tackled in science. Particularly, graduate students and their supervisors are encouraged to follow this approach as risk-mitigation plan to face the endeavors of PhD projects.

The case studies carried out so far were evaluated only qualitatively. However, the participants reported an increase in productivity, a higher success rate in the achievement of project deliverables, and, thanks to the enhanced number of contact moments, an improved perception of the working environment. More importantly, all the involved participants confirmed to be willing to further adopt *Agile* in the upcoming research challenges that they will be confronted with.

The promising results from finalized and ongoing case studies in different research domains encourage the establishment of additional collaborations with other Academic institutions. These collaborations will aim at quantitatively evaluating an *Agile* management of research projects on a statistically relevant number of subjects.

Acknowledgments

The authors would like to thank L. Comegna and D. Otyuskaya for the useful discussions in the initial stage of this work, and the members of the CaRE group at the Laboratory for Chemical Technology (University of Ghent) for being the first 'guinea pigs' of the proposed approach.

The authors are particularly grateful to all the participants of the currently ongoing seven case studies for their trust, time and dedication to the project.

References

1. C. Woolston, Graduate Survey: a Love-Hurt Relationship, Nature (2017)
2. K. Levecque, F. Anseel, A. De Beuckelaer, J. Van der Heyden, L. Gisle, Work Organization and Mental Health Problems in PhD Students, Research Policy (2017)
3. L. Constantin, How to handle the one-size-fits-all PhD, Nature Career Column (2018)
4. C.D. Tupper, Data Organization Practices, Data Architecture: from Zen to Reality, Morgan Kaufmann (2011)
5. H.L. Gantt, Work, wages and profit, Engineering Magazine (1910)
6. X. Amatriain Rubio, https://www.slideshare.net/xamat/agile-science (2008)
7. L. May and T. Runyon, LabScrum: A Case Study For Agility in Academic Research Labs, PsyArXiv (2019)
8. O. Bergman, https://crastina.se/agile-project-management-taught-to-students-at-karolinska-institutet/ (2019)
9. Manifesto for Agile Software Development (2001)
10. J. Sutherland, SCRUM The Art of Doing Twice the Work in Half the Time, Random House Business Books (2015)
11. Agile Project Management: Quick Start Guide, Albany (2016)
12. L. Pirro, How agile project management can work for your research, Nature Career Column (2019)
13. M. Montero Carrero, Ready to try Agile for research? Join our upcoming study, Ristretto.black (2019)

569949 TA Survival Guide: Navigating Your First Week of Classes

The first day walking into the class room as a teaching assistant can be scary, but with the right preparation, you can be successful in the classroom. There are a few steps that you can take to make your first week of classes as painless as possible. This can include talking to your instructor to negotiate your role in the classroom, understanding the tools available to you, and learning how to gain confidence and rapport with your students. In this workshop, you will learn how to navigate all of these obstacles and how to successfully ensure that you will have a productive and successful semester as a TA. You will be able to meet your developmental goals, be an asset to your instructor, and be a proponent for student success.

580029 Tunde Ogunnaike: Teacher, Author, Researcher, Administrator

This paper covers Tunde Ogunnaike's career in industry and academia and focuses on his contributions to education, research, and academic administration.

564194 Polymer Reaction Engineering in the Origins of Life:How to Get from Synthetic Rubber to Chemical Evolution

How does a chemical engineer go from synthetic rubber production to origins of life research and still be in polymer reaction engineering? The answer is simple: There are significant problems in chemical evolution (and likely in many other fields of research) where the tools of polymer reaction materials can be applied very effectively. This talk documents two of them.

In the study of chemical evolution, there are two schools of though. The first, often called *the RNA world*, RNA developed before proteins, and hence before enzyme catalysts. There is evidence for this in the fact that some polynucleic acids (RNA or RNA-like) function even today in an enzymatic capacity. If RNA was formed in the *RNA world,* then it must have been capable of replicating, not as it currently does by the use of polymerase enzymes, but without enzyme function, essentially by self-assembly. In the first part of this talk, methods of polymer reaction engineering were used to develop a mechanistic model of nonenzymatic polynucleotide replication. The model is made possible by the rate constants developed by Harvard researchers. This model treats the ideas of stalling proposed by the Harvard team in a direct kinetic mechanism. When one assumes that multiple incorrect insertions in a row are unlikely, the model returns estimates of the error threshold very much in agreement with the modified Eigen model (based on probability arguments) proposed by the Harvard team. In addition, the current model gives average lengths of polynucleotides, categorized by ultimate nucleoside (type and correct/incorrect) as well as the average lengths of interior and terminal sequences of correct and incorrect insertions in a row.

In the second view of chemical evolution, proteins were formed first, and eventually developed the enzymatic capability that enables essentially all of the chemistry of modern life. Although it is generally accepted that amino acids were present on the prebiotic Earth, the mechanism to form long chain polypeptides is still unclear. The second part of the talk, will discuss work in the noncatalytic formation of long chain polypeptides. We have found a simple system capable of forming peptide bonds under mild conditions. Hydroxy acids form metastable oligoesters in the oscillating (hot dry/cool wet) system and are converted into mixed copolymers via the ester-amide exchange reaction. Here, we use HPLC coupled with UV/MASS SPEC to track the growth of oligomers from valine/ lactic acid mixtures. The copolymerization is simulated by our kinetic (population balance) model to validate the proposed mechanism and optimize feed policies that result in longer oligomers. Finally, we evaluate feed policies that result in longer oligomers containing only one hydroxy acid residue at the N-terminus of a polypeptide. Peptide polymerization by this route follows kinetics similar to those of living polymerization. These results suggest a prebiotically-plausible route to the synthesis of longer depsipeptides that might lead to the formation of secondary structures, and ultimately, enzyme catalysts.

580030 Estimating the State of a Dynamical System: From Kalman Filtering to Moving Horizon State Estimation

This talk presents an overview of techniques for estimating the state of a dynamical system from the system output measurements.

The Kalman filter represents one of the early triumphs of modern state estimation for linear systems. The conditional probability of the state given the time sequence of measurements is shown to be a normal distribution for a linear system with normally distributed and independent measurement and process noises. Maximizing this conditional density produces a convenient recursion for the mean (the optimal state estimate) and the covariance of the distribution.

When it comes to deriving system properties, however, such as the stability of an estimator, the statistical properties provided by the Kalman filter recursion are of little use. The fact that a statistically optimal estimator is not necessarily even a stable estimator shows the divide between these two approaches to state estimation.

We then rederive the Kalman filter as an optimization of a cost function, and show that analysis of the optimality properties enable us to derive a close relative of a Lyapunov function for state estimation that provides a stability and robustness analysis.

The talk concludes by presenting some open conjectures and open research problems in state estimation.

580028 Biological Inspiration for Network Synchronization

Inspired by the collaboration with Dr. Ogunnaike while I was a young post-doc at DuPont, my lab has undertaken a path towards discovering the basic principles that are responsible for the mechanisms that allow biological networks to operate in a highly efficient manner in challenging environments. The discovery of such principles requires a significant research effort in the construction of models for complex biological networks, the development of analysis tools for such networks, and the discovery of structures and architectures for the design and control of such networks. In addition, understanding of the underlying physiological control systems can lead naturally to therapeutic interventions, such as the artificial pancreas.

In this talk I will focus on the work that our lab has done to develop both novel analysis and simulation tools to probe biological oscillators, in particular, the ubiquitous twenty-four-hour circadian rhythms. Using these tools, we have been able to extract design principles of biological networks, with the foremost among these being the principles responsible for robust synchronization of multicellular networks. Over the course of the last several years, our group made progress in the design of algorithms that are guided by these principles and are applied to dynamic sensor networks. Time synchronization has proven to be critical in sensor fusion applications where the time of arrival is utilized as a decision variable. Herein, we present the application of pulse-coupled synchronization to an acoustic event detection system based on a wireless sensor network.

580198 Deep Knowledge Versus Deep Learning

Developing mathematical models that causally (deterministically) map measured process inputs to critical response variables (i.e., outputs) that depend on these inputs is acute to optimum process operations. This objective is coined "process modeling" for the purpose of this work. Due to advancements in sensor technology, the number of process variables and sampling frequencies have increased considerably in the last decade. Big data advancements in process (i.e., plant) data have created renewed interests in empirically-based process modeling, i.e., the use of model structures that have no theoretical basis but a form that allows accurate mapping of measured inputs into the response space. Real applications of empirical modeling have established high risks modeling inaccuracies when going outside the input region used to develop the fitted response behavior, i.e., extrapolation. An empirical approach that has gained renewed popularity is artificial neural networks (ANN). Researchers have contended that big data have potentially expanded the successful applications of "Deep Structured Learning" also known as "Deep Learning (DL)" (Schmidhuber, 2015, LeCun, et al., 2015). The basic hypothesis of DL is that with very large data sets, an ANN structure can be widely expanded (i.e., increasing the number of nodes and hidden layers) and this will lead to more successful process modeling. This hypothesis has been tested on simulated processes, which are quite limited in representing critical conditions of real processes. These limitations include highly correlated inputs, unmeasured disturbances, measured inputs that do not affect the response but are significantly correlated with inputs that do, among other real modeling input characteristics. Thus, to establish the effectiveness and claims of DL ANN, its modeling must be done on real chemical processes of large multiple-input, complex, and dynamic response behavior, where the model is shown to be affective over a prolong period of time for a particular process.

In Rollins et al. (2015), a "deep knowledge (DK)" modeling approach was applied to a pilot distillation column with nine (9) measured inputs. This work defines DK as the use of highly structured input mapping to the response space that uses differential equations with physically based theoretical structures evidenced by having model parameters with physically interpretable understanding with the ability to set mathematical constraints on them based on well-established first principle dynamic modeling understanding. In the work, the response is the top tray temperature, and a Wiener modeling approach is used that gives excellent test results on eight (8) "freely existing" data sets over a three-year period. A "freely existing" data set means that no effort was made to intelligently change the input variables based on an optimal experimental design methodology. Actually, Rollins et al. arbitrarily selected all the data sets from this column's historical database that contained data sets created by undergraduate chemical engineering students learning how to run the column or running the column for data collection for a lab course in unit operations. The DK methodology of Rollins et al. also embraces a deliberate approach for maximizing information content and input causality based on a comprehensive Jacobian Matrix analysis to identify model weaknesses in the nature of input relationships. The Rollins et al. Wiener results of eight test cases represent the challenge for DL ANN.

This work applies DL ANN to the eight test cases and the results are fourth coming. In addition, for comparison, and to improve ANN modeling, this work applies a principal component neural network (PCNN) methodology that the authors developed successfully in real data study modeling of critical physical properties using asphalt core samples (Ghasemi, et al., 2018a, b). Ghasemi, et al. developed this methodology to fit ANN structures using orthogonal inputs in a DK approach to maximize information in a Jacobian matrix fashion. This is the first application of PCNN known to the authors of this work.

1. Schmidhuber, Jürgen. "Deep learning in neural networks: An overview," Neural Networks 61 (2015), 85-117.
2. LeCun, Yann, Yoshua Bengio, and Geoffrey Hinton. "Deep learning," Nature 521.7553 (2015), 436.
3. Rollins, D. K., A. K. Roggendorf, Y. Khor, Y. Mei, P. Lee and S. Loveland, "Dynamic Modeling With Correlated Inputs: Theory, Method and Experimental Demonstration," Ind. Eng. Chem. Res. 2015, 54(7), 2136-2144.
4. Ghasemi, P., M. Aslani, D. K. Rollins, R. C. Williams and V. R. Schaefer, "Modeling Rutting Susceptibility of Asphalt Pavement Using Principal Component Pseudo Inputs in Regression and

Neural Networks," International Journal of Pavement Research and Technology, https://doi.org/10.1016/j.ijprt.2018.01.003.
5. Ghasemi, P., M. Aslani, D. K. Rollins and R. C. Williams, "Principal component analysis-based predictive modeling and optimization of permanent deformation in asphalt pavement: elimination of correlated inputs and extrapolation in modeling," Journal of Structural and Multidisciplinary Optimization, 2018, 1-19.

580031 Enhancement of Chemical Engineering Education through Design Thinking: Integration of Theory and Cyber-Assisted Methods

In the engineering discipline, it is of utmost importance to give value to applied learning because as engineers, we are expected to innovate, and innovations happen when theoretical ideas are implemented successfully. Design thinking is one such approach that can enhance the value of theoretical concepts and motivate the students to visualize their ideas in more meaningful ways. The five stages of design thinking include: (i) Empathize, (ii) Define, (iii) Ideate, (iv) Prototype, and (v) Test [1]. Traditional engineering curriculum focuses on the last four stages and thus leads to a skewed perspective among students with regards to problem definition, formulation, and solution. Furthermore, research in chemical engineering education has indicated the need for revisions in the traditional curriculum and teaching methods to meet the needs of a modern chemical engineer [2]–[4]. Utilizing data analysis, programming and simulation tools is one way of demonstrating the virtual implementation of ideas and validating them at low costs and minimum risks. Thus, I have developed cyber-assisted modules as a part of the 'Process Dynamics and Controls', 'Process Optimization' and 'Engineering Clinic' courses offered to the undergraduates in the chemical engineering department at Rowan University, which strive to encompass all the five stages of Design Thinking.

For each topic in the Process Dynamics and Control, Process Optimization and Engineering Clinic courses, a supporting computer lab module was designed to substantiate student learning and simultaneously enhancing their mathematical modeling and computational skills. The lab sessions included a tutorial and an in-class computational assignment consisting of carefully chosen examples to closely resemble the state-of-the-art problems in chemical engineering. The tutorial consisted of an example from each topic, solved using analytical and computational (Matlab, Simulink, GAMS (General Algebraic Modeling Systems), R, Excel, etc.) methods. This provided the students with an opportunity to compare multiple solution platforms and choose the best to their liking and ease-of-implementation of the problem they intended to solve. Allowing students to work on the in-class assignments problems and discussing their computing issues, provided additional insights into learning barriers than traditional lecture-based teaching. We were able to discuss problem formulation errors, incorrect solver or algorithm selection, programming syntax errors, illogical initial guesses, lack of an output block or function, illogical results generated from ill-conditioned problems, etc.

Due to this integrated training, towards the end of these courses, the students were able to formulate and solve large-scale, real-world case studies and problems with broader contexts. Examples of problems solved by the students in team projects include drug dosing policies for chemotherapy treatment, an air-conditioning system in a classroom, configuration of perovskite crystals for harnessing solar energy, refractive-index matching of hydrogels, skiing trail-map optimization to maximize travel efficiency, etc. This integrated effort created a sense of appreciation for the course topics among the students and their application skills were considerably enhanced. They were able to think about the problem from multiple perspectives as well as implement their ideas virtually through cyber-assisted methods.

Keywords: computational tools, design, process control, optimization

References:
[1] R. Wolniak, "The Design Thinking method and its stages," Systemy Wspomagania w Inżynierii Produkcji, vol. Vol. 6, iss. 6, 2017.
[2] T. F. Edgar, B. A. Ogunnaike, J. J. Downs, K. R. Muske, and B. W. Bequette, "Renovating the undergraduate process control course," Computers & Chemical Engineering, vol. 30, no. 10, pp. 1749–

1762, Sep. 2006.
[3] B. W. Bequette and B. A. Ogunnaike, "Chemical process control education and practice," IEEE Control Systems, vol. 21, no. 2, pp. 10–17, Apr. 2001.
[4] T. F. Edgar, B. A. Ogunnaike, and K. R. Muske, "A global view of graduate process control education," Computers & Chemical Engineering, vol. 30, no. 10, pp. 1763–1774, Sep. 2006.
[5] I. Luka, "Design Thinking in Pedagogy," The Journal of Education, Culture, and Society, no. 2, pp. 63–74, 2014.

568156 The Manifest Fingerprints of Excellence in Education and Mentoring

This presentation will chronicle my interactions with Tunde Ogunnaike throughout my career, from a 1st year graduate student to now an Associate Professor. I will in particular focus on how his insatiable drive for excellence in all forms of scholarship, without compromising integrity in personal interactions, has shaped my science, lab and life. He is a master communicator--both written and oral--and this underlies much of his excellence in both teaching and research. We should all be learning from and striving to his examples.

581102 The Myriad Service Opportunities in ChE Teaching and Learning: Find One for You

The goal of this presentation is to encourage you to find a way to serve in chemical engineering teaching and learning. I will discuss some of the many service positions I have held. I'll also mention extensions of those positions, all to spark your interest and help you find one that fits you. I will discuss benefits, pitfalls, and perceived obstacles of serving our community, as well as ways to avoid the pitfalls and overcome the obstacles. We will have time for questions about opportunities for you.

577590 The Integration of Visualization into Education through Comics and Animation

One of the major obstacles to effective learning in STEM is the abstract nature of many core STEM concepts. While traditional approaches include examples of real-world applications, demonstrations, and/or hands-on experimentation, the depth and complexity of these concepts can cause such efforts to be unsuccessful. Further exacerbating the problem, some theoretical concepts do not have obvious visual representations that could make the concepts less abstract and so more comprehensible to students.

The benefits of broadening visual instruction in education are not just restricted to learners with particular inclinations toward visual learning. Studies have shown that combining text with images improves students' retention of information, and multimedia presentations of concepts allow for better transfer of the gained knowledge towards solving problems. In short, increasing the visual content of STEM instruction promotes students' conceptual understanding.

Recently, the use of comics and animation as learning tools has grown dramatically, with a broad set of artists, book series, and videos all dedicated to integrating visualization into educational approaches. This recent growth stands in stark contrast to the long history of comics as educational tools, during which the known potential of comics in learning has fought a societal perception of comics that has limited research and impact over much of the last century.

This presentation will focus on the range of visual techniques being used; the potential of comics and animation in improving student enthusiasm, confidence, and understanding; and the impact of these techniques observed thus far.

581101 "Do I Need to Know This for the Exam?" Shifting Students' Engagement Towards Professional Practice

Some have mused that if Rip van Winkle were to wake up today, the only thing that would look familiar is the educational system. That may not be entirely true. There has been emphasis recently in transitioning chemical engineering classroom instruction from transmission-based lectures to active learning. However, even in active learning classrooms, it is not uncommon to find students who do well on exams but cannot operationalize that same material in a practical project. Conversely, in our research, we have observed students who have performed poorly in the traditional classroom but excel in team activities that position them as engineers undertaking meaningful tasks. These observations raise a fundamental question about student engagement in the active learning classroom: Engagement to what end? We design engaged instruction to shift students from "doing school" to acting like professional engineers, and have developed supporting technology that draws students into cooperative and critical participation, instead of sequestering them in front of computer screens. In these learning environments, student interactions with one another and with instructors reflect the professional engineering workplace in which identifying and solving problems is a collaborative process that requires use of engineering principles in a dynamic sense-making environment. Importantly, providing space for students to share multiple perspectives and bring their own valued real world experiences creates opportunity for a more inclusive classroom. My presentation examines "reasoning episodes" in which students working in teams collaboratively argue from evidence, construct explanations, and formulate and revise models as they design processes. We argue that such real-world learning environments best enable students to develop more connected conceptual understanding as well as develop needed professional skills to better prepare them to meet the engineering challenges and civic responsibilities of the 21st century.

581100 Centering a Community: Building an Aiche Education Division

In 2006, Professor Michael Cutlip asked a casual question to the presenter, “Why isn’t there an Education Division in AIChE?” That question initiated a series of events that ultimately transformed an idea into an entity that now functions as a model organization within AIChE. This talk will describe the process of developing the Education Division, with a focus on the approach taken by the many individuals who contributed to creating the Education Division and laying the foundations of the thriving community that it is today.

581103 Education Division: Present and Future

An open forum will be held for attendees to ask questions about the current state and future directions of the Education Division. Elected, volunteer leaders who serve as officers of the Education Division will be available to answer questions regarding the division. Your comments and questions will help lead the discussion.

576517 Risk Assessment in Unit Operations Laboratories

Incorporating chemical process safety into the Unit Operations Laboratories is the goal of six faculty and instructors from six different universities. Seven specific learning outcomes are being integrated into the curriculum, as recommended by the AICHE SAChE Committee to meet ABET requirements. One of these learning outcomes is to introduce the process of hazard evaluation and risk assessment.

Risk assessment entails identifying the frequency and the consequence of an incident. In order to identify the frequency of incidents in UO labs, two different incident reporting structures were implemented in two different universities. These reporting structures allow students, teaching assistants and faculty to report incident frequencies. Reports generated from these structures enable the programs to quantify the frequency of incident occurrences.

The other aspect of risk assessment is identifying the consequence. There are a range of consequences and these consequences must be categorized in terms of low, medium and high. While death and millions of dollars in damage may be obvious high consequences, the tolerance for risk may be much lower in an academic environment, For instance, a high consequence incident may be a student burn or damage to a $100 piece of equipment. Utilizing the collective knowledge of the group, a consequence scale was established.

This presentation will present details on the reporting structures and a risk matrix that may be used for university unit operations laboratories.

552669 Using Simulators for a Competitive Process Safety Exercise

Columbia University Chemical Engineering and Simulation Solutions Inc. have developed a competitive process safety chemical engineering exercise using simulators in order to educate students on chemical engineering process safety. As part of a four-day Chemical Engineering Operations Center educational exercise combining computer simulation and classroom instruction this team educates students on chemical engineering process safety and provides a unique chemical engineering process safety experience. In 2019, this academic and industry team developed a new method of using the simulators inspired by Esports. Multiple years of integrating simulation of chemical engineering processes, the CHEN E4501 Chemical Engineering Process Safety Course uses case studies, computer aided design, and simulation in the course content covering instructional objectives. The results of the 2019 lesson sequence and the Competitive Process Safety Exercise sponsored by Simulation Solutions Inc. and the methods used to evaluate student performance as related to safety outcomes are reported on here. This presentation will focus on the competitive exercise results.

578885 Classroom Use of Process Simulators to Quantify Flashing Two Phase Release Conditions

Process simulators are a readily available tool for making complex thermodynamic calculations which are necessary for quantifying the conditions for flashing two-phase flow under sudden depressurization. Many common toxic industrial chemicals (TICs) such as chlorine and ammonia are stored as liquids at ambient temperature, and these materials will flash when depressurized to atmospheric pressure forming an aerosol. This paper considers how to use a process simulator to estimate the mass flow rate from a simple hole for flashing two-phase flow. Previously published data are considered to demonstrate the effectiveness of this approach. The result of this work is the development of a general approach for quantifying the mass release rate of flashing two phase releases for TICs and other flashing liquids including mixtures with supercritical components which cannot use current textbook methods.

557652 Chemical Engineering Student Moral Reasoning within Hypothetical Process Safety Contexts

In light of recent industrial process safety incidents, ABET requires accredited undergraduate chemical engineering programs to train students to make effective and accurate process safety decisions. While many programs train their students to effectively identify and evaluate risks relevant to process safety, the continued occurrence of incidents indicates that elements of the environment in which engineers are tasked with making decisions may not be effectively captured through current instructional methods. As such, we have examined how senior chemical engineering students apply moral reasoning when faced with process safety decisions in two contexts: a text-based assessment instrument and a virtual process safety environment.

In the text-based assessment, students took the Engineering Process Safety Research Instrument (EPSRI) which consists of 5-7 process safety dilemmas and decisions. Students are then asked to rate the importance of 10-12 considerations, unique to each dilemma, to their decision-making process. In the second method, a virtual environment was developed to simulate a chemical plant, including safety hazards, interactions with co-workers, and other authentic constraints. After certain decision points, students were asked to rate how important certain considerations were to those decisions. Based on the responses to the considerations on the EPSRI and the virtual environment, students' moral reasoning could be determined by the ratings they assigned to considerations that were pre-conventional (focused on self), conventional (focused on immediate contacts such as family, employees, and company), and post-conventional (focused on the greater community) as outlined in Kohlberg's moral development theory. Comparisons of students' moral reasoning between the text-based EPSRI and more authentic environment of the virtual simulation will be discussed in the context of behavioral ethics, examining how the different scenarios impact ethical behavior and decision-making.

553646 The 4th Edition of Chemical Process Safety, Fundamentals with Applications

It has been almost 30 years since the first edition of *Chemical Process Safety, Fundamentals with Applications* was published in 1990. The second edition was published in 2002 and the third edition in 2012. The fourth edition will be published in May of 2019 and is a major revision - several chapters were completely rewritten and almost all chapters heavily revised.

This paper will discuss the original intent and history of the textbook. In addition, it will present details on the 4th edition, including new content that was added and content that was removed. We believe the 4th edition realigns the textbook with current process safety concepts and practices and provides an even more meaningful learning experience for engineering undergraduates.

567960 Reading Analytics and Cramming Using an Interactive Textbook for Material and Energy Balances

Big data was generated using an interactive textbook resulting in almost 300,000 reading interactions for three annual cohorts. Specifically, the Material and Energy Balances zyBook has numerous features where reading clicks and homework answers are compiled for students' and instructors' use – including earning a small portion of the final course grade. Previously, median reading rates were observed as high as 99% for over 1,200 recorded clicks per student. However, additional data is available for reading after each assignments' due date to compare to previous results when reading assignments were due. When examining reading analytics after the semester, additional reading was measured despite no incentive of earning a course grade. In total, up to four cohorts representing over 380 students will be presented. Conclusions drawn from the end of semester data will be compared with previous, reproducible findings, including A and B students read statistically significantly more than C+D+F students and female students reading more than male students.

568743 Improving a Graduate Pedagogy Course to Support Inclusive Teaching and Active Learning

New chemical engineering graduate students at UC Berkeley are required to take a one-semester course that introduces them to the theory and application of teaching and learning within the discipline. This training happens during or before their first term as a graduate student instructor (teaching assistant), and the course contributes in important ways to the development of the teaching and learning philosophies of graduate students in the Chemical and Biomolecular Engineering Department. The impacts of the teaching and mentoring skills of these graduate students are felt throughout the undergraduate body of around 450-500 undergraduate students, and continue in academic and professional careers of the graduate students.

This poster details work to update materials in the course, especially given recent research data on the effectiveness of active learning strategies for conceptual idea development in chemical engineering, and given new information on inclusive teaching and mentoring to support learning for a diverse student body. Materials that address ideas of intersectionality of race and gender, mentoring within and across various identities, and awareness of structural inequities and cultural biases in higher education are timely and necessary for the continued improvement of teaching practices within our department. The work includes engagement of national experts in chemical engineering education, and work of a graduate research assistant to review and update materials in collaboration with the faculty member. Changes to the course are implemented in the fall 2019 semester. Graduate student essays in the course and responses on course surveys are used to evaluate the effectiveness of course improvements.

570520 It's Electric! Incorporating Battery Experiments into Senior Unit Operations

Senior unit operation courses often focus on traditional chemical engineering units that are learned from earlier fundamental courses. However, in a dynamic world where chemical engineers are working in a broader range of industries than ever before it is imperative to adapt the senior unit operations lab to reflect the changing landscape for future chemical engineers by incorporating more non-traditional experiments in senior unit operations. One new experiment that is feasible to adopt in most laboratories are battery experiments. The copper-zinc and aluminum-air batteries are two economic and safe batteries that can be adopted into lab and provide an enriching experience to students. During these experiments students can alter variables such as reagent amounts, surface area of metals, electrolyte composition and concentration.

571951 Strategies for Hands-on Activities Integrated in an Undergraduate Engineering Curriculum

It has been demonstrated that thehuman body is one of the most efficient chemical factories able to produce chemicals of complex nature that help maintain body structure and function. Understanding the complexity of individual cellular and molecular machines, and how their high hierarchical system integration leads to proper body functionality will help unravel how and why adaptability is based on changing environmental and societal conditions.

Within the context of human body complexity and functionality we proposed to provide physical means for visualization of body-related functions in synthetic environments to be used in curriculum development strategies in biomedical engineering. Specifically, we will discuss two strategies, namely designing an engineered-based system that mimics the transport in biological systems and building educational models using 3 D printing techniques to increase understanding of engineering-related concepts of biological systems respectively.

We demonstrate how hands on activities integrated in the curriculum not only leads to better learning but further, how the challenges and future prospects for hands-on activities implementation could be used as a pipeline for dissemination of engineering related concepts and techniques to high school students to thus provide concrete examples of real-world integration and applications while allowing exposure to STEM disciplines.

575054 Design of an Open-Ended Unit Operations Experiment in Transport Phenomena

The junior unit operations course at UCSB focuses on experiments in thermodynamics and transport phenomena. Students typically rotate through four 2-week experiments over the 10 week quarter. While the experimental rotation gives students exposure to many types of experiments, there is limited time for students to explore areas of interest outside of predetermined experiments. This open-ended experiment was designed to reinforce concepts in momentum and heat transfer and build upon their problem solving skills by allowing the students to explore an area of interest. The experiment consists of a variable-speed, centrifugal pump mounted to a work stations with an exchangeable discharge pipe that allows for many types of experiments to be rapidly "dropped-in". Students are able to choose from several proven experiments including the measurement of frictional losses in pipes, characterization of multiple types flow meters, measurement of fluid velocity profiles, and development of pump performance curves. In addition, students are able to rapidly manufacture prototype parts via 3D printing to test new part designs that otherwise would be cost or time prohibitive to acquire.

USE OF BUSINESS MODEL CANVAS IN BIOPROCESSES ENGINEERING EDUCATION AT UNDERGRADUATION LEVEL: ENCOURAGING ENTREPRENEURSHIP AND INCREASING MOTIVATION

Ângela M. Moraes, Monize C. Decarli, Cecília B. Westin, Marco C. P. Soares, Franciele F. Vit, Fernan D. M. Jimenez and Lucimara G. de la Torre

Department of Engineering of Materials and of Bioprocesses, School of Chemical Engineering, University of Campinas, Campinas - SP, Brazil. E-mail: ammoraes@unicamp.br

Abstract

In the university environment, entrepreneurial practices and motivational stimuli can be efficiently interconnected, further enticing students to achieve success in performing their activities. With the purpose to consolidate contents of the area of bioprocessing, to promote the training of students in entrepreneurship, as well as to increase their motivation within the classroom, it was applied, in a compulsory course of the Chemical Engineering undergraduation program of the University of Campinas, in Brazil, an evaluation activity based on the development of a Business Model Canvas. The focus was on the industrial production of a product based on a bioprocess defined by the students, covering concepts of enzyme kinetics, kinetics of cell growth, substrate consumption and product formation, design of bioreactors for operation in batch, fed-batch, continuous and perfusion modes, mass and momentum transfer effects in bioprocesses, and downstream processing. The approach not only played an important role in the consolidation of knowledge in the bioprocesses area but also contributed to the formation of dynamic and communicative professionals, offering them a better groundwork to face the competitive job market.

Introduction

The need for an increase in the number of students engaged in science and technology, combined with a demand of professionals experienced in team-working and used to the human interactions commonly observed on both office and industrial processes motivated the pursuit of educational innovations [1,2]. In more recent years, such innovations have been contributing to the formation of professionals able to perform professionally to better attend to the more complex needs of the modern society [3-5]. The innovations studies are mainly focused on two dimensions of the teaching and learning process: 1) a wide dimension, that consists in the reevaluation of the curricular structures [6]; 2) a localized dimension, focused on the implementation of the teaching techniques, to enhance the students' comprehension and learning efficiency [7,8].

The studies suggested that intuitive and active learning methodologies are efficient for the enhancement of students' motivation and engagement. These methodologies consist on the presentation of a specific challenge by the instructor, and the students must discuss and recognize their own lack of knowledge for performing the task, getting intrinsic motivation, while the teacher is supposed to only guide the students through the process of obtaining the knowledge by themselves, instead of just solving problems [9]. In fact, many educational papers claim that the engagement of the students in the processes of comprehending the phenomena substantially enhances their understanding of the problems and situations, making the learning process more effective when compared to the traditional expositive lectures [1,10].

A way that could help teachers to apply that is through the use of tools such as the Business Model Canvas (BMC), proposed by Osterwalder and Pigneur in 2010 [11]. This tool helps in structuring thinking, facilitating learning, being a concise tool that is usually used to assist the rapid and efficient formulation of several business models, from mapping existing business models and designing new models to managing a portfolio of business models. Another relevant tool is the Lean Canvas, which has been proposed by Maurya in 2012 [12], a less detailed approach primarily directed to entrepreneurs and startup businesses focusing the most risk-associated aspects of creating a new business.

Basically, the BMC is subdivided into blocks (Figure 1), which should be filled with direct statements. No particular order is recommended to fill the nine BMC building blocks. All elements identified as relevant to the model should be initially listed and their positions in the canvas can be moved to better describe the problem focused. BMC completeness and coherence should be checked mainly to detect the presence of isolated elements, e.g. key resource and activities not associated to value propositions.

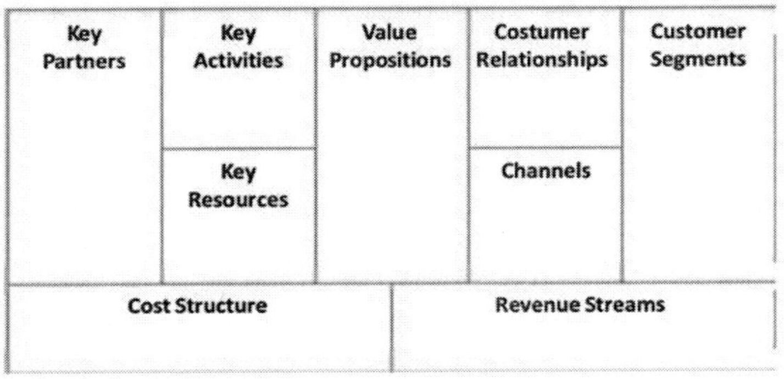

Figure 1 - Summarized block structure of the business model canvas [adapted from [11]).

The BMC analysis strategy is based in simple and straightforward concepts to describe, design, challenge, invent, and pivot the business model, which stimulates the students. Based on that, and in order to keep the students motivated to deal with more dynamic academic and industrial environments [1], a methodology based on the effective interconnection between entrepreneurial practices and motivational stimuli was proposed in this work and successfully applied to the traditional program of the Bioprocess Engineering Course (EQ883), a compulsory topic of the Chemical Engineering program of the University of Campinas (UNICAMP), in Brazil. This work had, therefore, the purpose to consolidate contents of the area of bioprocessing, to promote the training of students in entrepreneurship and critical analysis, as well as had the perspective of increasing the motivation of students within the classroom.

The students were then graded in this course not only through the use of traditional written exams, but also according to their performance in the development of a Business Model Canvas. The focus was on the industrial production of a product based on a bioprocess defined by the students themselves. Concepts of enzyme kinetics, kinetics of cell growth, substrate consumption and product formation, design of bioreactors for operation in batch, fed-batch, continuous or perfusion modes, mass and momentum transfer effects in bioprocesses, and of downstream processing and post-production processing aspects were involved in all proposals.

Methodology

The proposal to create the Business Model Canvas (BMC) was carried out by students from 6 classes of different semesters divided into teams. Each team focused on the traditional topics of the BMC for each type of product, from product design and biocatalyst-mediated transformation to final product purification. Thus, in addition to the technical aspects inherent to each process, value propositions, key partners, key activities, key resources, customer segments, customer relationships, distribution channels, cost structure, and revenue streams had to be presented for each case.

In all classes, the presentation of the results of the work occurred in stages, at least two of which were in the form of seminars given to the rest of the class. The first step was always the oral presentation of the company, product or service selected, as well as the motivation for this choice, lasting 5 minutes. The intermediate stages involved in-depth discussion of the results directly with those responsible for the course, and at the end of the academic semester the consolidated results of each team were presented orally not only to the instructors and the rest of the class, but also to a panel of external evaluators, who played the role of investors who compared the different investment proposals in a comparative manner.

The evaluation criteria were based on the following aspects: a) relevance of the chosen product; b) content, depth, and relation of the case studied with the main objective of course; c) technical approach adopted for the development of each stage of the production process; d) market and economical aspects; e) development of the Business Model Canvas; f) organization and integration among team members; g) duration of the seminars, based on pre-established periods; h) visual resources used in the presentation; i) team commitment to written and oral communication; j) analysis of the critical view of the team regarding the other teams' presentations.

The evaluation process is summarized in Figure 2. In the first stage, the technical committee that evaluated the proposals was formed by the responsible teacher, the teaching assistants assigned to the course, a master's or PhD student and a postdoctoral researcher (Technical Commission 1). In the final stage, the technical committee was formed by the teacher, the assistants and by external members, such as a professor from another undergraduation program at UNICAMP or from another university, entrepreneurial partners from animal cell culture and biomaterials segments, members of entrepreneurship consulting companies or from INOVA, the technology transfer office of UNICAMP (Technical Commission 2). In the first stage, all students and members of Technical Commission 1 could invest in the proposals of at least two teams considered as the most attractive (the students could not invest in their own company). In this step, the effect of variation in the companies' stock values was also investigated. In the second stage, only the members of Technical Commission 2 invested, and after a consolidation meeting, the winning company was announced.

Results and Discussion

The number of students in each class and the topics chosen by them for the BMC proposals are listed in Table 1. The average grades of the students in this activity as well as their overall grades and the weight of the BMC in the final grades are shown in Table 2.

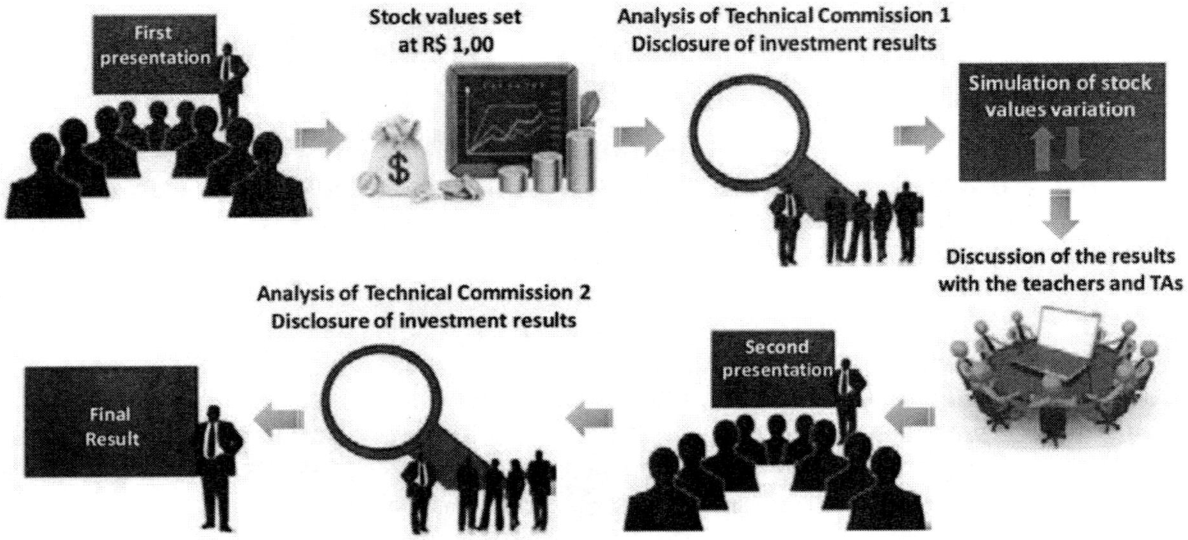

Figure 2 - Flowchart of the proposed work.

As observed in Table 1, 275 students so far benefit from the use of BMC tool in EQ883 at UNICAMP. A wide range of topics were covered along the years, from the production of beer and other beverages to the manufacture of products to be used for biomedical, cosmetic, animal feed, fuel and cleaning purposes, among others.

The first two times the BMC activity was applied (2nd semester of 2016 and 1st semester of 2017), groups with significantly higher number of students were formed in comparison to those of the other semesters. At those semesters, the groups were divided in three sub teams (each with 4 to 6 students), one focused on market aspects, the other in process kinetics and reactor design and the last one in downstream and post-production processing. This approach demanded extra team work, since the students had to coordinate and share their assumptions, tasks and calculations, but it was observed that there was a loss of vision of the overall project because of the large sizes of the groups, so this organization model was no longer used.

One of the times (during the first semester of 2018), two classes on the same EQ883 topic were taught simultaneously by two teachers, however many of the tasks and works were shared among all the students, including the BMC proposal activity. At this particular time, groups mixing students from both classes were formed, and all presentations were performed for all students, of both classes. A difference of less than 2% was noticed for the average grade in the BMC activity and of less than 1% for the average final grade between the two classes.

The data shown in Table 2 also show that the BMC proposal activity was an important component of the grade of the students, being significantly higher (from 14 to 30%) than their average global grades.

Relevant aspects of the proposed activity were identified. Initially, the selection of the product of interest notably generated enthusiastic discussions in the students, as it allowed them to actively choose, in a free-way, a product whose production they could undertake in the future. In many cases, there was also a need for field research on the requirements of the product to meet the demand and also its applicability.

Table 1 - Number of students enrolled in EQ883 Bioprocess Engineering course classes from 2016 to 2019 and the topics chosen to be developed by them.

Class (semester/year)	Number of students	Products chosen by the teams
1st/2019	43	1. Biodiesel production through lipase catalyzed transesterification 2. Biogas from corn residues 3. Ethanol obtained from residues of paper industry 4. High fructose syrup 5. Lipase production for incorporation in powdered detergents 6. Surfactants (ramnolipids) production by *Pseudomonas aeruginosa* 7. Yogurt with no added sugars or thickeners
2nd/2018	27	1. *Aloe vera* components extraction, hydrolysis and processing for incorporation in hygiene products 2. Biosynthesis of vanillin 3. Buthanol produced through IBE (isopropanol-buthanol-ethanol) fermentation 4. Insulin 5. Laundry enzymes 6. Lipase production for the treatment of dairy industry effluents 7. Wound dressings containing papain for lesion debridation
1st/2018	Class A: 27 Class B: 37 Total: 64	1. Buthanol produced through IBE (isopropanol-buthanol-ethanol) fermentation 2. Craft pilsen beer 3. Gluten-free beer 4. Hyaluronic acid 5. Keratinase for the hydrolysis of chicken feathers usable in animal feed formulation 6. Lactase 7. Lactic acid 8. Lactose-free milk 9. Lipase 10. Wine with reduced alcohol content 11. Xylitol
1st/2017	39	1. Aloe vera components extraction and processing for incorporation in cosmetic products 2. Gluten-free beer 3. Sodium gluconate extracted from figs
2nd/2016	59	1. Craft beer 2. Immobilized lactase 3. Insulin 4. Probiotic cheese

Students were encouraged not only to seek information about the characteristics of the production processes but also to find kinetic parameters and to define appropriate operating conditions to meet the proposed goals. They also had to get acquainted to contents associated with the opening of a business.

An example of the final BMC presented by one of the teams is shown in Figure 3. It describes the business model of Insulbras, a phantasy-company created for the production, research, development and commercialization of insulin obtained from genetically modified *Escherichia coli* fermentation.

Table 2 - Average grades of the students in the BMC proposals, their overall grades and the percentual composition of the BMC activity in the student grades.

Class (semester/year)	**Weight of the BMC activity in the final grade (%)**	**Average grade in the BMC activity**	**Average final grade**
1st/2019	20	8.6	7.2
2nd/2018	20	7.9	6.9
1st/2018 - class A	20	7.9	6.8
- class B	20	8.2	6.9
1st/2017	15	8.3	6.7
2nd/2016	15	8.7	6.7

It is possible to observe in Figure 3 the summarized contents of the nine blocks that constitutes the BMC, showing the remarkable capacity of the students to put together a rather complex process in the form of a business proposal throughout the 4-month EQ883 Bioprocess Engineering course. The Value Proposition, in the center, describes the problem that would be solved by the company (high-quality insulin production with competitive costs) and the reason why Insulbras' products are more valuable than the ones from competitors (the company was projected for the improvement of peoples' life quality through innovation). The Customer Segment identifies the groups of people to be reached and served by the addressed product (in this case, patients with diabetes). The Channels are the ways the products are supposed to be distributed, how the customers will be reached by Insulbras and who will perform the direct contact with them (drug stores and logistic companies). The Customer Relationships define the relations between the company and its clients and how they will be maintained (Insulbras would maintain a direct contact with medical doctors and pharmacists and a customer service). The Key Resources refers to the most significant assets required for the success of the proposed business model (Insulbras needs high quality raw materials for the insulin production, *e.g.*, *E. coli* cells and culture medium, and man force for driving both the industrial and business processes). The Key Activities (control of production, quality and distribution, as well as research and development) are the ways Insulbras is supposed to transform its Key Resources into final products and services. The Key Partners or Partner Network include the external financial, logistic and other business works that provide resources or activities that Insulbras depends on, but are not on its core business; therefore, it would not be viable to maintain such activities among the internal processes, or the company simply does not have the expertise to perform them (Insulbras depends on the distribution and use of insulin by hospitals and drug stores, which have medical and pharmaceutical professionals allowed to administrate the drugs, and on the suppliers who transport and deliver the products under appropriate conditions). Financial aspects represented by the Cost Structure, involve the costs for the development of the business, which must be aligned with the Value Proposition and the main ideas of the Canvas (Insulbras must spend with raw materials and vessels for the drugs, with the implantation and maintenance of an infrastructure adequate for production, distribution, research and development of insulin, with its labor force, governmental taxes and permits). Finally, the Revenue Streams, which is associated to the value the customers are willing to pay to obtain the product and represent how the company expects to earn money, is presented (Insulbras sales are focused on insulin flasks).

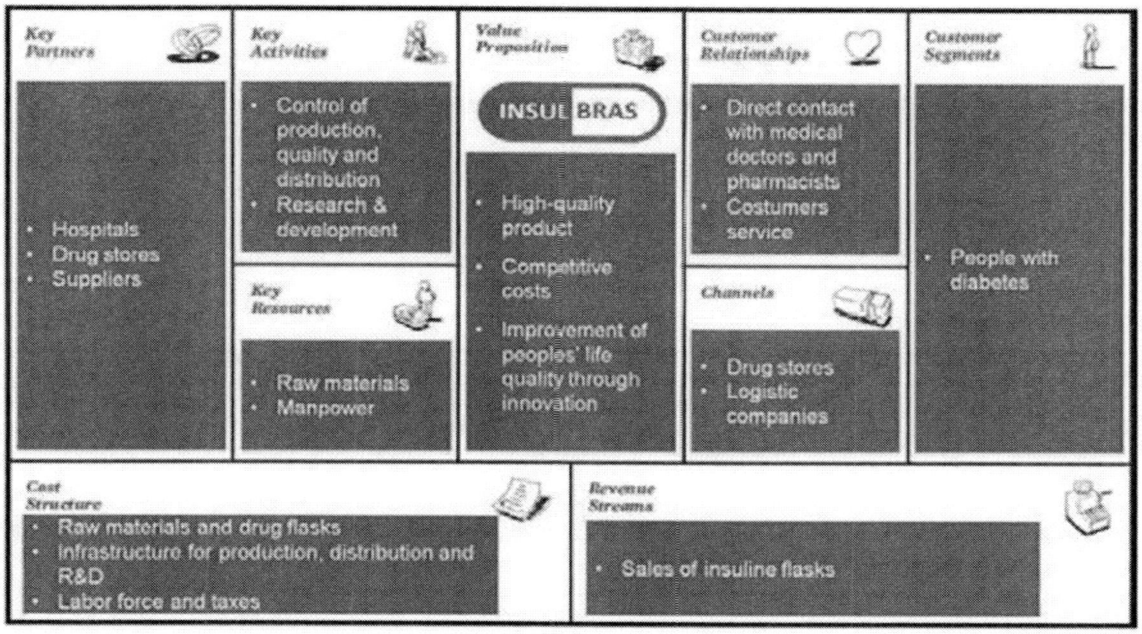

Figure 3 – Business Model Canvas elaborated by students for an insulin production company named Insulbras.

All these topics were extensively discussed by the groups' members to result in the final canvas and as frequently during the presentations assumptions made by one team were questioned by other groups, student critical analysis was also significantly exercised.

Even though none of the nine individual elements of the BMC is new for business people, it is an excellent way to exercise the capacity of summarizing the whole business logical structure of a company in a single-page document. This simple, but holistic view of a business model, has been continuously developed and improved by thousands of users in the last decade and the fact that it is freely available certainly contributes significantly for that.

Conclusions

It was observed with this work that the introduction of entrepreneurship concepts in courses directed to Chemical Engineering students of undergraduate level allows them to anticipate an important real-work experience, with consequent development of the capacity of analyzing the business environment and professional expertise requirements. Furthermore, this activity not only played an important role in the consolidation of knowledge in the bioprocesses area but also contributed to the formation of more dynamic and communicative professionals, with a broader view, and of chemical engineers with a much more significant entrepreneurial vision, offering them a better groundwork to face the competitive job market.

Acknowledgments

The authors would like to acknowledge the support to this research by the National Council for Scientific and Technological Development (Conselho Nacional de Desenvolvimento Científico e Tecnológico – CNPq), the Coordination for the Improvement of Higher Educational Personnel (Coordenação de Aperfeiçoamento de Pessoal de Nível

Superior – CAPES – Finance Code 001), the São Paulo Research Foundation (FAPESP) and the Teaching, Research and Extension Support Fund from UNICAMP (FAEPEX), all from Brazil. The authors also acknowledge the contribution of Ariane Baylon Dias.

References

1. Dym, C.L.; Agogino, A.M.; Eris, O.; Frey, D.D.; and L.J. Leifer (2005), "Engineering Design Thinking, Teaching, and Learning", J. Eng. Educ., v.94, pp.103–120.
2. Godfrey, E. (2009), "Exploring the culture of engineering education: The journey", Australas. J. Eng. Educ., v.15, pp.1–12, 2009.
3. King, R. (2008), "Engineers for the future: Addressing the supply and quality of engineering graduates for the 21st century", Sydney: Australian Council of Engineering Deans with support from Australian Learning and Teaching Council.
4. Jamieson, L.H. and J.R. Lohmann (Eds.) (2009), "Creating a culture for scholarly and systematic innovation in engineering education: Ensuring U.S. engineering has the right people with the right talent for a global society", Washington, DC: American Society for Engineering Education.
5. Spalter-Roth, R.; Fortenberry, N.; and B. Lovitts (2007) "The acceptance and diffusion of innovation: A cross-disciplinary approach to instructional and curricular change in engineering", Washington, DC: American Sociological Association.
6. Graham, R (2012), "Achieving excellence in engineering education: the ingredients of successful change", The Royal Academy of Engineering & Massachusetts Institute of Technology, Report, March 2012.
7. Crouch, C.H and Mazur, E. (2001), "Peer Instruction: Ten years of experience and results", Am. J. Phys., v.69, pp.970-976.
8. Apple, T. and Cutler, A. (1999), "The Rensselaer Studio General Chemistry Course", J. Chem. Educ., v. 76, pp.662-463.
9. Prince, M. and Felder, R. (2007), "The Many Faces of Inductive Teaching and Learning", J. Coll. Sci. Teach., v.36.
10. Hake, R.R. (1998), "Interactive-engagement versus traditional methods: A six-thousand-student survey of mechanics test data for introductory physics courses", Am. J. Phys., v.66, n.64.
11. Osterwalder, A., & Pigneur, Y. (2010) "Business model generation a handbook for visionaries, game changers, and challengers". New York: Wiley.
12. Maurya, A. (2012). "Running lean: iterate from plan A to a plan that works". Beijing: OReilly.

576002 Laying a Foundation with Outcome Assessment in Material and Energy Balances

Most chemical engineering programs assess achievement at high levels of learning across the palette of ABET student outcomes in capstone courses such as senior design. This poster will demonstrate a similar assessment, albeit at low levels of learning, in a sophomore-year material and energy balance course. A primary objective of the approach is to familiarize students with the expectations of a professional engineering graduate, using the framework of ABET student outcomes. In doing so, it also circumvents the "Why do we have to do/know that?" question often posed regarding those assignments later in the curriculum that address the development of professional skills such as teamwork and communication.

Of course, problem-solving is the most fundamental learning objective for any material and energy balances course. Here, we will focus instead on supplemental assignments addressing ethical responsibility, experimental data, written communication, team function, independent learning, and basic design choices. The assignments, presented to students with an explanation of why those skills are important, invite them early on to embrace the development of the knowledge, skills, and behaviors associated with an engineering professional, thus laying the foundation that will ultimately support the capstone experience.

576060 Integrating VBA-Excel into the Curriculum across Seven Junior-Year Courses

In many chemical engineering curricula, the junior year is where students first encounter engineering problems that require iterative or numerical solutions, or problems that, although solvable by hand, are much more tractable using computational tools. In years past, departments required formal programming courses in FORTRAN, C++ or other languages to enable students to formulate solutions to complex problems. Programming has become less common in the chemical engineering curriculum as departments now rely on high-level tools such as MATLAB and MathCad.

However, the increasing cost of these high-level tools and insistence from industrial stakeholders that they will not be readily available to graduates has led some departments to adopt Visual Basic for Applications (VBA) in Excel as the preferred programming tool. Microsoft Excel is ubiquitous in industry, and VBA is a common tool for many potential employers; additionally, students are already familiar with Excel from first year engineering courses. VBA-Excel offers both the versatility of a programming language and the ease of a spreadsheet.

Rather than create a full 3-credit hour programming course, we introduced VBA across the curriculum in the junior year spanning seven courses. Student are not only taught the fundamentals of VBA, but also develop tools to solve complex problems in a variety of topics from thermodynamics to reactor design using the language. In this work we will demonstrate how VBA is introduced in a fall semester computational lab course and employed in three fundamental core courses contemporaneously and in three additional courses the following spring.

576322 Fair or Foul – Food Freshness Beyond the Smell Test

Packaging is a topic that is not often discussed in the undergraduate chemical engineering curriculum. In practice, it is generally something that is overlooked since the contents are often what the end-user cares most about, and packaging the barrier to get to it. However, packaging in the food and beverage industry is particularly important because manufacturers have to ensure that the (often) perishable contents are able to withstand the physical and thermal stresses that the product may encounter from plant to consumer.

In this work, we will ask undergraduate chemical engineering students in a heat transfer class to: 1) identify a commercial food or beverage product they wish to study, 2) research and obtain properties of its packaging (e.g. paper, aluminum, cellophane), 3) research spoilage data on the food product, and 4) model the heat transfer through the packaging over several real-world thermal-stress scenarios using Comsol to identify when the product will spoil. Some examples of these scenarios include: 1) an inadvertently long transfer from delivery truck to the grocery store refrigerator on a summer day, 2) power outage, and 3) a stress scenario of the student's choosing. Additionally, students will be asked to attempt to improve the food packaging they studied, with the goal of extending the viable timeframe of the product. Possible avenues of continuation for this project in the chemical engineering curriculum include the mass transfer course (e.g. modelling the effects of leaks in the packaging), as well as the kinetics or biotechnology course (e.g. examining the actual growth kinetics of the bacteria).

559288 Unit Conversions Meet Google Analytics: Insights into How Chemical Engineers Use Mobile Apps

This presentation will detail highlights of usage of functionality in mobile apps designed by and for chemical engineers as anonymously collected via embedded Google Analytics over a 6+ year span with over 1,000,000 data points. Of particular interest are the categories and pairs of units which are most (and least) often converted, which may offer insights into how engineers study, learn and work. Furthermore, such knowledge can help guide and streamline teaching in Mass & Energy Balance courses, as the data suggest that a somewhat limited set of units and unit conversions are of real and practical interest. Additional data sets and insights relating to chemical properties and other types of routine chemical engineering calculations will also be presented.

569745 Homework Problem Repository and Delivery System for Quantitative Problems

Student engagement with course material is key for student comprehension. One traditional method for promoting this engagement has been homework assignments. This system, however, has been degraded over the last few years due to the availability of homework solutions. A freely available, easy to use, crowd sourced, homework problem repository available at QRproblems.org/login has been developed to help address this degradation.

The repository has many features meant to make it attractive as a resource for fellow instructors. The repository can be sorted on course, concept, author, difficulty, computational method or contributor. Each problem that has been previously circulated shows the students rating of the problem in terms of effectiveness and difficulty. Currently there are hundreds of problems, in the repository with more added each semester. Some problems are based on examples from text books or online resources, while others are original. A problem must be activated by an instructor for students to access it. Only students that enter their PIN, instructor ID and problem number can access an active problem. This should reduce the exposure of the repository to the general public and make it somewhat more difficult for problems solutions to be generally available. Another feature of the repository is the ability of instructors to clone a problem. If an instructor likes a problem but wants to make modifications, the problem can be cloned and resubmitted as a cloned problem to the repository.

All problems have variable parameters so that each student gets a different version of the problem. Students get immediate feedback on their numerical responses either on their phones or computer. Students are also provided the answers but not solution to a common base-case version of the problem. Instructors can then insist that students attempt to work through the common base-case problem before asking for help. Many of the problems have additional enhancements such as video clips, simulations or solution hints in the form of multiple-choice questions. Problems also have written reflection questions that the instructor can assign that supports higher level thinking and attempts to connect the concepts in the problem to reality. In addition, planning or estimation preliminary problems can be assigned some time before the numerical problem is due. Completion of the preliminary problems is automatically logged by the system. This feature was included to encourage students that have not developed time management skills to read and understand the problem statements several days before the problem is due.

How fellow instructors can use and contribute to the repository will be demonstrated.

575235 Simview VR: a Software Pipeline for Virtual Reality Real-Time Molecular Dynamics Simulation and Visualization of Arbitrary Molecules

Simview VR: a Software Pipeline for Virtual Reality Real-Time Molecular Dynamics Simulation and Visualization of Arbitrary Molecules

Heta A. Gandhi, Rainier Barrett, Andrew D. White

3D visualization of molecular simulations has gained attention over the last few years. Virtual reality (VR) 3D visualization toolkits generally involve bulky equipment and difficult-to-use interfaces. We have developed a VR visualization application which takes in a SMILES string along with simulation parameters and renders the molecular dynamics (MD) simulation trajectory on a VR headset in real-time. All simulations are done in the HOOMD-blue simulation engine, and the graphics for the VR are rendered in Unity. The key features that set this software apart from previous 3D viewer programs are real-time feedback and ease of use. This application can be used for real-time diagnostics of MD simulations for research purposes, and as an interactive tool for educational purposes.

566617 Predicting Success in Undergraduate Chemical Engineering at the United States Military Academy

Standardized college entrance examinations, such as the SAT or ACT, are used as admissions criterion at many colleges and universities in the United States. Standardized entrance examination scores not only differentiate applicants, but several studies have shown that examination scores can predict future academic success. Our results suggest a positive correlation between College Entrance Examination Rank (CEER), a preadmissions tool created and used by the U.S. Military Academy at West Point, as a predictor for successful graduation and success on the Fundamentals of Engineering Exam (FEE). Thus, we propose that a similar pre-admission criterion such as CEER can be useful to other universities when considering engineering program applicants. At the U.S. Military Academy at West Point, all cadets are required to excel not only in academics, but also physically and militarily. Performance in each area is continually assessed and cadets are given an aggregated cadet performance score (academics accounts for 55%, military 30%, and physical 15%) each academic term. A cadet's aggregated performance score can help determine future opportunities in the Army post-graduation, such as branch selection (e.g., infantry, aviation) and post selection (e.g., Fort Carson, Colorado or Fort Benning, Georgia). Failure to maintain minimum threshold standards in each of the three aforementioned areas can lead to separation from West Point. To predict future success in each area, West Point created the Whole Candidate Score to assess cadet candidates during the admission process, which includes an academic subcomponent called the CEER. The CEER includes not just SAT or ACT scores, but other metrics, such as high school academic rank. Cadets can select majors across 15 different departments including, chemical, civil and mechanical engineering as well as history and philosophy. We chose to focus our efforts on analyzing the performance of chemical engineers, a major requiring two additional courses compared to the rest of the student body and a heavy emphasis on complex engineering and mathematical content. Our study examines whether a pre-admission predictor of successful graduation could be applied to further predict success within an individual program. Further, we postulated whether this predictor could not only apply to successful graduation but to performance on the nationally administered Fundamentals of Engineering Exam (FEE).

571757 A Look into How Students Take up Realistic Instructional Tasks

One common critique of the engineering curriculum is that students leave unprepared to connect the knowledge they learned in the classroom to the messy, open-ended work they face in engineering practice. The study described in this talk was part of a broader institutional change initiative where we are attempting to address this issue by creating assignments that better resemble the work done in engineering practice in 11 core studio courses. These group activities cast students in the role of professional engineers working on design projects, allowing multiple solution paths. We want students to learn practices that will be useful to them as engineers. Practices are learned within contexts and through the use of PBL, our intent is to shift the context in which students learn foundational concepts.

In a qualitative case study, we analyzed the activity of three teams that were video recorded as they made progress on one such authentic task. Being a laboratory study, the students were taken out of the context of worrying about a grade. However, two of the teams initially responded using the types of practices and reasoning that corresponded to doing well in a class but do not directly transfer to engineering work. Adapting Dorothy Holland's construct of figured worlds, we label this state of engagement as being in school world. On the other hand, one team's activity reasonably coincided with engineering practice; we say they were engaged in engineering world.

When you have students do PBL, you ask students to navigate between these figured worlds. However, much of the students' experiences in engineering programs resides in school world. In this study, we give them a PBL problem designed to have students engineering world thinking and examine what they do with it. We present two distinct forms of engagement, school world and engineering world. In this talk, we compare the activity, processes, interactions, and thinking in each figured world and show ways that engineering world is preferable to school world in better preparing students for practice

We also look at the ways activity from school world to engineering world evolves evolve and shift in this context and what we can learn about designing curricula to place students in engineering world. The two teams that showed similar patterns of beginning in school world wound up shifting into engineering world over the course of the activity. Shifts into engineering world are preceded by multiple unsuccessful bids by one team member who we label the engineering world bidder. We argue that it is not be a single event that shifts the team from school to engineering world, but rather a history of the team in activity, and that the failed bids contribute to building of dissatisfaction with school world practices. When they encounter a "snag" that school world practices cannot resolve, the team, as a collective, changed activity to engineering world. We also see a shift in participation from school world where a single high-status student directs the team's activity to engineering world where participation shifts and is shared more evenly throughout the group.

572920 An Interdisciplinary Research Site for Emerging Researchers within the Functional Materials and Manufacturing Institute at University of South Florida

Early research experiences would benefit from an environment of reinforcement and practice of basic skills within a supportive network, which also provides guidance towards career options. Such an environment could also provide a variety of opportunities to develop and practice communication skills, and provide exposure to more challenging topics such a design and integration. Undergraduate research experiences in such an environment can be expected to increase retention of participants in research careers, especially via pursuit of graduate education. High school experiences in such a program are expected to increase interest in STEM careers. The Functional Materials and Manufacturing Institute (FMMI) at University of South Florida (USF) engages in interdisciplinary materials and engineering research, and its outreach programs explicitly aim to provide such an environment to diverse participants, which include high school students, REU program participants, international interns, and high school teachers. Functional materials are prevalent throughout the modern world, and are constant reminders of advances from engineering. Research experiences in functional materials, with a mindset toward application/manufacturing, is an outstanding topic to engage emerging researchers with a wide variety of interests. FMMI research experiences are designed to include substantial time towards hands-on research, while also emphasizing non-research elements including career guidance, technical writing and presentation skills, safety, ethics and team-building. In this contribution, the overall program for emerging researchers at FMMI will be discussed with an emphasis on best practices for integrating multi-tiered efforts in an interdisciplinary environment. Research highlights and achievements from the participants will also be presented.

571991 Developing an Undergraduate Research Opportunities Programme (UROP) to Improve Students' Professional and Research Skills

Many engineering students in the UK do not get real exposure to scientific research until the last year of their undergraduate degree or during their Master's degree. This means it will be difficult for them to know whether they would like to pursue a PhD or research career until very late and PhD deadlines have closed. In addition, the research offered during their degrees are often not very challenging projects offering little novelty and no chance of being a co-author of a publication. Similarly, many students do not have contacts in industry or possess the skills needed to get a position.

Therefore, Imperial College has introduced something called an Undergraduate Research Opportunities Programme, a 10-12 week summer programme open to undergraduates from any year, where the students get to work together with a research group on a real, often industrial research topic. Students can apply for a stipend similar to what they would receive if taking up an internship in industry to support them while doing their UROP.

Seeing the success at Imperial College, I opened up this programme to students from other universities that did not offer the same opportunities for their undergraduates. In order to alert students about the opportunity I contacted various departments at different universities in London and the interest was overwhelming. In order to promote teamwork and collaboration between disciplines, a team of students from a diverse range of educational backgrounds were selected. In addition, to promote equality and diversity, the gender balance and ethnicity were also taken into consideration to give applicants an equal opportunity.

The cohort of students started the first few days with research and safety inductions. Afterwards, the students had to solve a number of problems in teams, both experimental as well as computational, which were closely related to their research projects and present their findings. This was where the importance of different educational background was crucial - they had to help each other in order to successfully solve the problems. After the initial training, they started on their individual research projects, many of them in collaboration with industry, which often involved updates and presentations to the sponsor. During the duration of the programme they had to give presentations to a variety of audiences as well as submitting short update reports and finally journal paper draft.

Many of the previous UROP participants have had their research published as well as continued careers in related fields, both as PhD students and R&D Scientists in industry. Past participants have often kept in contact with our research group over the years and told us about how useful their experience was, which proved that the programme was successful in giving the students both the relevant professional and research skills.

571762 Resource Development Approach for Enhancing Enabling Environments within Experiential Learning Modules

Resource development approach for enhancing enabling environments within experiential learning modules

Umang V. Shah, Pavan Inguva, Clemens Brechtelsbauer

[1] Department of Chemical Engineering, Imperial College London, South Kensington Campus, London SW7 2AZ, United Kingdom

Introduction

With the rapid emergence of new technologies in the 21st century, chemical engineers have to tackle challenges that are often multi-disciplinary. This regularly forces them to obtain new knowledge beyond their zone of expertise in order to remain competent and relevant. From the educational perspective, this means institutes of higher education should not only prepare their students for *independent learning* but also for *independent thinking* and *creative problem solving*.

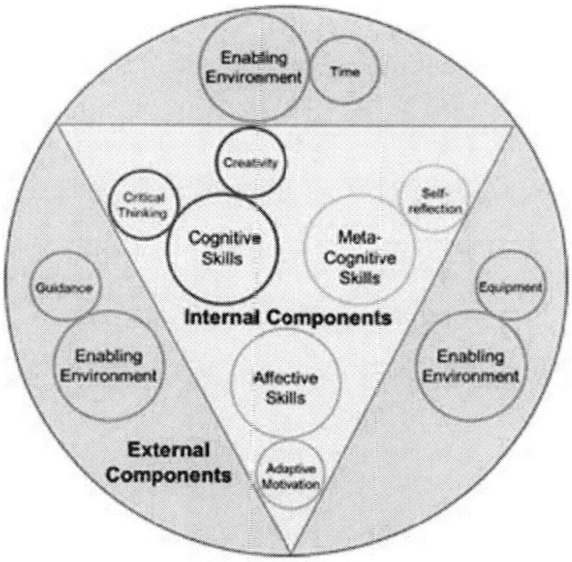

Figure-1 Necessary components for independent learning from learner perspective

Independent learning has several internal and external components with respect to the learner

(Cukurova, 2014) (**Figure 1**). The internal components include cognitive, metacognitive and affective skills of the learner, whereas the external component refers to an "enabling environment" that includes facility, time, peer and teacher/facilitator.

Internal components for promoting independent learning have been discussed at length in the literature. On the contrary, external components are rarely discussed. Considering the focus on educational literature on lecture based modules/ problem classes, external components may be less critical, however if independent learning is to be promoted in a hands-on education, the external components become equally important, particularly equipment and support infrastructure.

This study focuses on the pedagogic underpinning of how sound resource design can facilitate effective experiential learning and ultimately promote independent and cross-disciplinary learning using a case study of a 2nd Year chemical engineering laboratory module. We propose a resource design approach that consists of three pillars: 1) Flexibility, 2) Authenticity and 3) Accessibility (FAA).

Course Context

The Knowledge Lab is a practical module offered to second year chemical engineering students as part of their four year MEng degree. Knowledge Lab aims to develop a student's applied problem-solving skills and to provide a glimpse of a professional process engineering workflow. In this module, students pose as group of consultants to solve a real-life engineering problem and have the flexibility to design their own solution approach which has to be agreed with their project supervisor. Each week of work is termed 'a step' which increase in complexity over the three weeks duration of the project.

A typical example of the project set-up and its pedagogic goals

A typical example of a project offered in the course can be demonstrated by the reactor commissioning project, where student role play as commissioning engineers tasked with commissioning a set of reactors to produce a product at a target yield. The Process and Instrumentation Diagram (P&ID) and equipment is shown in Figure 2. The set-up is designed to enable students to design and implement control strategies for a variety of process parameters to optimize reactor performance for maximum conversion.

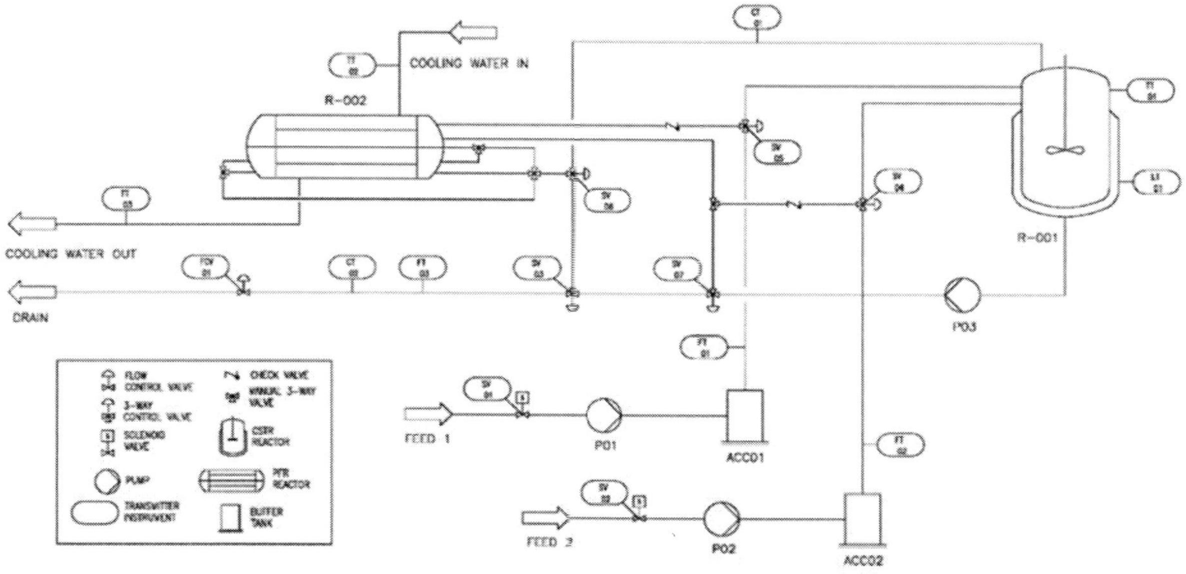

Figure-2 P&ID for reactor commissioning project

The external components have to meet three pedagogical design goals to facilitate independent learning:

1) Flexibility:

   a. Flexibility in project conception and design:

      i. The experimental design offers a high degree of operational freedom. From a control standpoint, students can develop different control strategies by selecting from multiple possible actuator and sensor options. In addition, students have flexibility in selecting what part of the project they want to focus on (such as optimizing reaction vs. developing control strategies). The focus of the approach is not on finding the correct answer, but on being able to understand and explain the received answer.

   b. Flexibility in assessment:

i. All submission deadlines and assessments are negotiated with the project leads, thus providing flexibility with deadlines.

ii. Considering the intellectual load, the majority of the assessment is focused on discussions of the decision rationale and explaining results in one to one conversation with the project leads, which not only provides instant feedback and intellectual discussion possibility.

2) Authenticity:

a. Experimental set-up: For any experimental set-up, parts and instrumentation are industrial grade. The control user interface is based on industrially accepted Supervisory Control and Data Acquisition (SCADA) principles (Figure-3). All resources are designed to represent a typical academic or industrial research lab.

b. Teaching material: The handouts were structured as project briefs similar to that given to practicing engineers. Students need to take initiative in making decisions on aspects such as data-processing and presentation which would be expected professionally.

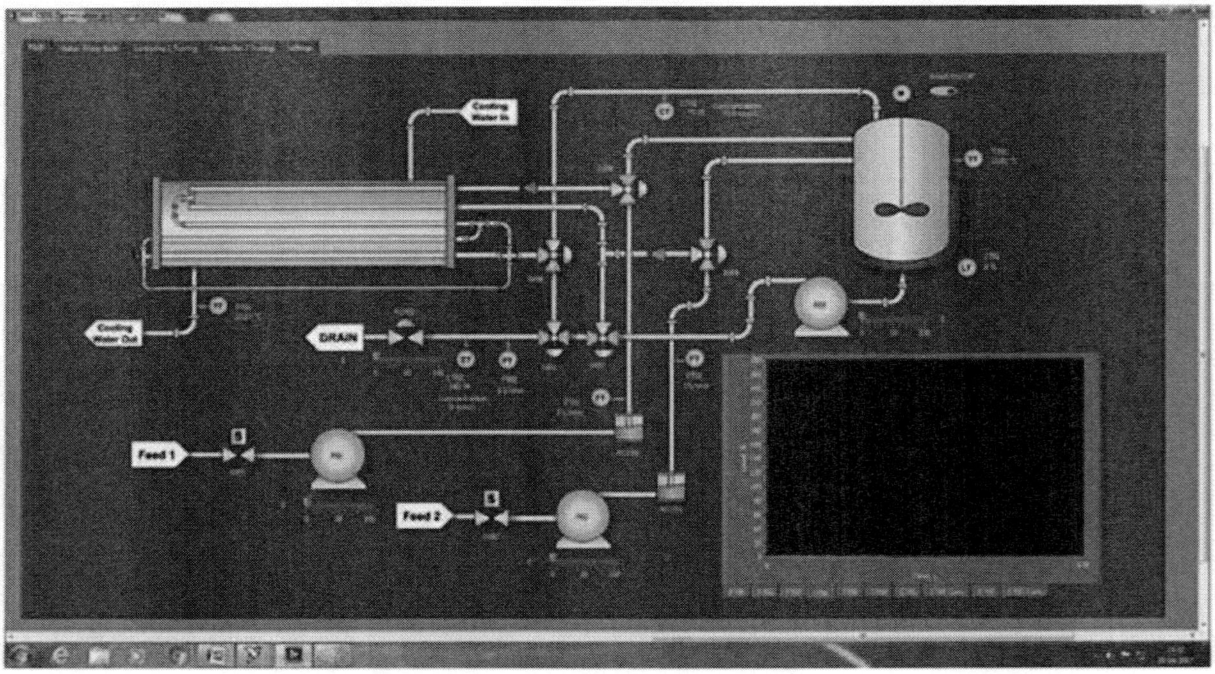

Figure 3 Control interface for the reactor commissioning project

3) Accessibility:

a. Experimental set up: All experimental set-ups are designed to be easily accessible to students, which means the design of the interface includes many user-friendly features for example. This helps to reduce time spent on equipment familiarization and operator confusion, thus facilitating a student's independent exploration of the design space. Additional resources, i.e. video led operational instructions, equipment and individual instrumentation manuals, calibration details, etc. are provided in addition to the brief project handouts.

b. Teaching material: To help facilitate student's independent exploration of available literature, a reference list of key articles and textbooks were provided.

Evaluation of Learner Outcomes

The resource design approach on the basis of FAA, has been on offer since autumn 2016 and three cohorts have worked with this approach.

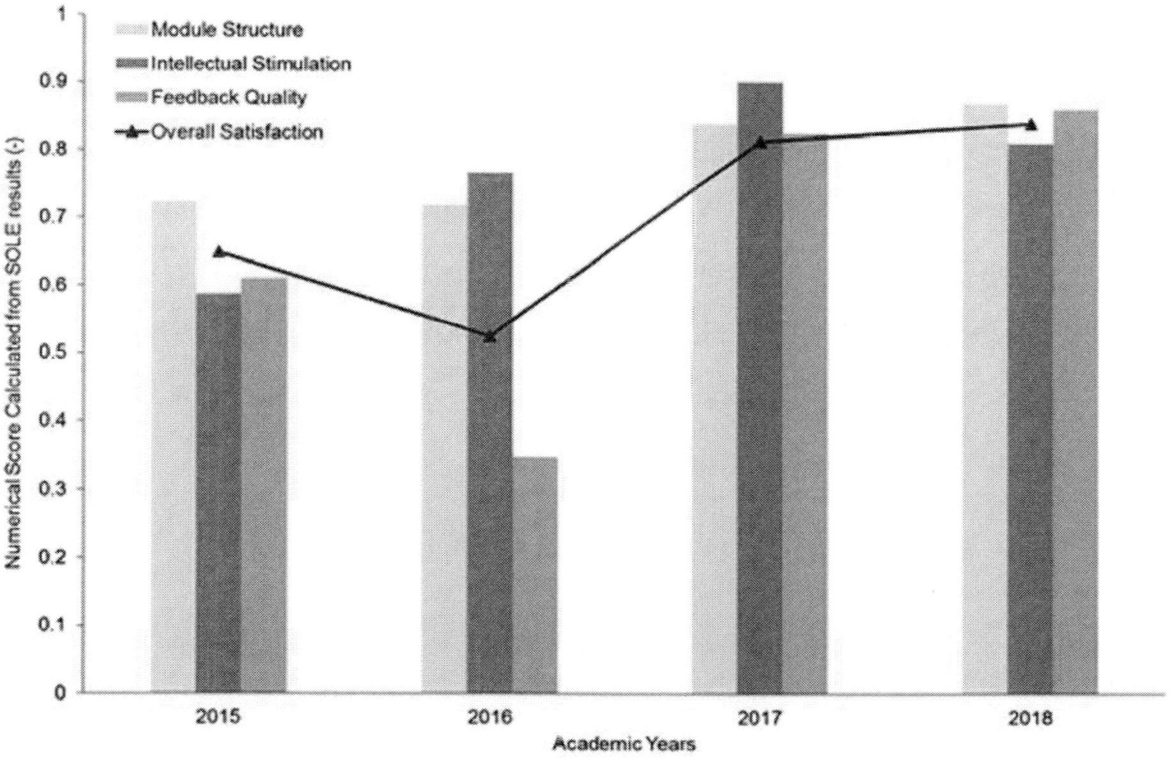

Figure 4 UG student feedback results for Knowledge Lab (2015-18)

Figure 4 shows student feedback on the module (n ³ 80), for each cohort commencing from autumn of the year on four different metrics. Feedback was collected and measured on a Likert scale. Evidently, student satisfaction in multiple modes has improved significantly with the introduction of the new Knowledge lab resource design approach. It is interesting to note that in 2016 which was the first year of the new structure, overall satisfaction dipped. The cause for this decline was noted in the free text comments to be primarily teething issues with the new equipment which have subsequently been fixed. None of the free text comments voiced any issues regards lack of engagement and intellectual stimulation.

In addition to the student feedback, focus group interviews were conducted for current and past 2nd year students and graduate teaching assistants. The purpose of the interview was to evaluate the effectiveness of the pedagogic approach.

Discussion

Focus group discussion revealed that flexibility, authenticity, and accessibility were found to be three important pillars in the experimental resources. As highlighted by one of the GTAs during the discussion, at the educator level, design and resource flexibility means the same project can be used to formulate a variety of hands-on activities targeting different student autonomy levels ranging from demonstrations to open ended projects. Within the same autonomy level, different activities can also be designed, varying in project aims, methods and materials, without compromising authenticity, and accessibility. At the teacher level (GTAs), they can adapt the equipment to meet the requirements or experimental plans as proposed by the students. At the student level, they have significant operational flexibility with exploring the design space.

Benefits of FAA for Students - Promoting Independent Learning

The large number of operational degrees of freedoms in set-up means that students have a large design space to work in. This approach enhances the learning cycle as follows:

Abstract Conceptualization: instead of working with a pre-existing rig configuration, students first have to consider the various possible permutations of the equipment and draw on previously learnt theory before deciding on a plan of action.

Active Experimentation: students can test various combinations of the equipment when setting up a control loop or testing different control strategies without spending too much time on adjusting the hardware or software.

Reflective Observation: students can quickly see the effect of their experimental choices due to the lab scale nature of the equipment, thus allowing them to reflect on the relationship between the settings used and the outcome

The accessibility of an experimental set up is important for creating an enabling work environment. This environment enhances the following aspects of Kolb's cycle.

Active Experimentation: An easy-to-use set-up and an adequate support mechanisms means that students are not intimidated by the rig or spend time getting it to work. Having fail-safes in terms of equipment configurations as well as in assessment gives students the reassurance that they can fully explore the design space for these experiments without risking failure or compromising marks, reducing perceived risk and mental barriers during experimentation. The transformation axis from active experimentation to reflective observance is also activated as the resource flexibility makes it easier for students to physically see the impact of their experimental choices, facilitating the reflective observation process.

Conclusion

The nature of the different components of enabling environment, including experimental resources, support infrastructure, and assessment can have a significant impact on learning outcomes and module development. Design flexibility, authenticity, and accessibility is evidently demonstrated to be key in enhancing the effectiveness of the enabling environment. This in turn enhanced the learning process through greater activation of all the aspects of experiential learning.

References

Cukurova, M. (2014) *An Investigation of an Independent Learning Approach in University Level Chemistry: The Effects on Students' Knowledge, Understanding and Intellectual Attributes*. University of York.

580106 A Perspective on Teaching Process Safety through the Curriculum

It is important to properly place process safety topics in the chemical engineering curriculum to emphasize the importance of the topic and to introduce students to the key concepts. One of the best approaches, in principle, is to cover chemical process safety topics in most, if not all, core courses in the curriculum. In this paper, we will discuss the pros and cons, as well as give example of how this might be accomplished.

574468 An Initiative on Safety across the Chemical Engineering Curriculum Scott Fogler

Sometimes chemical process safety is taught in a separate safety course within the chemical engineering curriculum, and sometimes it is taught only in the senior year as a part of the process design course. The purpose of this initiative is to provide faculty and students with real case studies and resources so that process safety can be more effectively and easily learned throughout the curriculum and become an integral part of chemical engineering culture.

To achieve this culture, safety modules have been developed to be used as homework problems in every core chemical engineering lecture course (http://umich.edu/~safeche/). Each module consists of viewing a Chemical Safety Board Video, filling out an analysis of the accident, doing a course specific calculation, assigning the NFPA symbols and filling out a Bow Tie Diagram for the accident. For example, in the chemical reaction engineering course, the T2 laboratories accident and the ExxonMobil Refinery Fire are examples presented and analyzed. In addition to the core lecture courses, six video snippets (2-5 minutes each) on Laboratory Safety are included on the website.

In addition to the modules, tutorials are given on such things as the NFPA diamond, the Bow Tie Diagram, the Fire Triangle, the Process Safety Triangle and A Safety Analysis of the Incident. The solutions to the homework problems are also given on the website, but instructors will have to email the author to receive the password to enter the solutions.

The website is free and accessible to all.Sometimes chemical process safety is taught in a separate safety course within the
chemical engineering curriculum, and sometimes it is taught only in the senior year as a part of the process design course. The purpose of this initiative is to provide faculty and students with real case studies and resources so that process safety can be more effectively and easily learned throughout the curriculum and become an integral part of chemical engineering culture.

To achieve this culture, safety modules have been developed to be used as homework problems in every core chemical engineering lecture course (http://umich.edu/~safeche/). Each module consists of viewing a Chemical Safety Board Video, filling out an analysis of the accident, doing a course specific calculation, assigning the NFPA symbols and filling out a Bow Tie Diagram for the accident. For example, in the chemical reaction engineering course, the T2 laboratories accident and the ExxonMobil Refinery Fire are examples presented and analyzed. In addition to the core lecture courses, six video snippets (2-5 minutes each) on Laboratory Safety are included on the website.

In addition to the modules, tutorials are given on such things as the NFPA diamond, the Bow Tie Diagram, the Fire Triangle, the Process Safety Triangle and A Safety Analysis of the Incident. The solutions to the homework problems are also given on the website, but instructors will have to email the author to receive the password to enter the solutions.

The website is free and accessible to all.

561465 Learning Hazards of Exothermic Reactions Via Real and Virtual Experiments

Although there is a recognized need for chemical reaction safety education, there is a dearth of published studies on experiments designed to demonstrate hazards of exothermic reactions. We developed an experiment to study scale-up of an exothermic reaction and provide students with dramatic first-hand knowledge of the hazards of thermal runaway. This experiment has been implemented successfully in our unit operations laboratory and in that of at least one other university. In this presentation we discuss a way for a broader audience of students to safely experience thermal runaway via a virtual experiment.

We developed a computer simulation for the iodide catalyzed decomposition of hydrogen peroxide using COMSOL Multiphysics® software. The simulation was converted to a user-friendly "app" that allows a scale-up study of this reaction by manipulating variables in a control panel and observing temperature and pressure profiles as the reaction proceeds under different conditions. Using COMSOL Server™ software we have provided access to the app to anyone with an internet connection and a web browser. The app provides a virtual experiment that requires no knowledge or installation of the COMSOL Multiphysics® software. It is anticipated that interacting with the app will provide students with a lasting understanding of the consequences of the fact that reaction rate, and concomitant temperature and pressure increases, scales with reactor volume while necessary heat removal typically scales with reactor surface area.

562663 Fundamental Approaches to Reactive Chemicals Hazard Recognition and Analysis

Over the past half century, Dow's Reactive Chemicals Program has supported a strong safety culture and a scientifically rigorous, risk-based review process for chemical hazards in the company's labs and plants. The foundation of Dow's chemical and process hazard analyses is the data generated by the Reactive Chemicals Group, who leverage measurement science, expertise in chemistry and engineering, and an understanding of Dow's processes to support safe operation at all scales. In spite of the critical role the Reactive Chemicals discipline plays in safe chemical research and manufacturing, these approaches and techniques are absent from the majority of chemistry and chemical engineering undergraduate curricula. In this talk, we discuss the fundamental principles behind assessing chemical reactivity hazards. We focus on material that could be incorporated into the undergraduate curriculum, such as assessing theoretical and experimental heats of reaction, calorimetric techniques for studying both desired and undesired reactions, and recognizing flammability and dust explosion hazards. Incorporating reactive chemicals testing or modeling into an undergraduate curriculum would satisfy expectations and requirements for safety instruction and better prepare students for careers in both industrial and academic roles.

578267 New Liquid Flammability Rating with the Assist of Machine Learning

Currently, the classification and rating of liquid's flammability is based on its flashpoint. However, flashpoint becomes insignificant when the liquid is present in the form of aerosol. This study uses DIPPR 801 database to find a new liquid flammability rating with the consideration of the possibility of atomization. Over 800 organic chemicals have been studied and principle component analysis is applied. Also, several unsupervised learning algorithms have been utilized to conduct the clustering on the database. The new liquid's flammability rating standard will extend the existing liquid flammability standards, such as NFPA 704 and GHS of Classification and Labelling of Chemicals.

578258 Practicing the ABCs of Technical Communication Using Realistic Problems and Homework Memos

Practicing solving problems through homework is an important aspect of building students' technical skill set. However, the format of these problems and the way they are reported do not generally align with the way problems are framed or reported on outside of the classroom. I have recently instituted a new type of homework problem where students are assigned realistic problems and instead of only turning in the handwritten or spreadsheet calculations, they must also submit a short memo reporting on their work in the context of a fictional company. While the memo format is not necessary, these style assignments give students practice with the ABCs of technical communication: Accuracy, Brevity, and Clarity. Through these memos, students learn to identify the key concepts and results from their work, communicate information concisely, and manipulate and present data in a manner that is accurate and intelligible. I will demonstrate example assignments from two classes: an elective on alternative energy and a core class on chemical engineering and the environment. In the alternative energy class, where some students had minimal technical writing experience, there was an observable improvement in written communication skills over the six memo assignments given that semester.

574688 Flipping Writing Revisions: Students Devise Self-Revision through Reflection in the Engineering Laboratory

A balance is yet to be struck between writing to learn and communication in the workplace across the engineering curriculum. Most engineering students find writing reports in core courses, including laboratory sequences, challenging or a task of reporting 'just the data', presenting a conundrum to instructors: How do we support engineering students to think deeply about core concepts and discipline-specific discourse simultaneously? Engineering faculty who teach laboratory often find grading to be burdensome, time-consuming, and unsustainable in the long-term. Furthermore, industry members want engineers who can communicate across different audiences in the corporate domain. This study focuses on a small group reflective writing activity designed with several goals in mind: sustainability and promoting critical self-assessment through reflection, a tool well-acknowledged as key to student learning and professional identity development. Participants include students (N=52) enrolled in a junior-level Chemical Engineering Laboratory II course at a large minority-serving research university in the American Southwest. A core English writing instructor was embedded in the course to assist students with written communication to a broader audience across the boundaries of the discipline. There were three major student deliverables in the laboratory: two individual short reports and one oral presentation per team. The reflection activity with each team was facilitated in the second round of short reports. We conducted mixed methods analysis on the student work. Individual student reflections were assessed by all the course instructors using the following criteria: indication of deep thought which translated into an audience-minded revision plan or sufficient argument against revision and specific evidence cited to support their perspective on revision. The activity was designed as a guided reflection on each section of the report reminding students that a report is a narrative, and the content of each section should advance that narrative. Higher scores were recorded for student reflections that included specific evidence from the report itself and direct language about readers or audience needs. Reflections that neglected to mention the reader's needs, or the writer's impact on the reader, and read like a to-do list received lower scores than reflections that mentioned one or more of the criteria. We found the reflections tended to fall into three categories: defense of rhetorical choices, fix errors like a to-do list, and audience-minded revision plans. Students' ability to have adaptable writing styles for different audiences was also evidenced by their citing of real world applications as broader context for an experiment in the report. The reflection activity provides students the opportunity to go beyond a typical checklist of report corrections to self-revising with audience in mind.

575419 Utilizing Peer Review to Improve the Quality of Unit Operations Assignments

After graduation, we quickly learn that technical communication requires a significant time commitment and many iterations on the same draft. Unfortunately, undergraduate students rarely have the time to go through the iterative process of perfecting their communication skills. This limits not only the quality of their assignments, but also their ability to learn successful strategies for improving communication. While allowing students the opportunity to revise communication assignments based on faculty feedback would be ideal, the huge time commitment associated with grading and providing feedback is limiting. Therefore, this presentation will focus on strategies for integration of peer review processes for both written and oral communication assignments into the unit operations lab. In going through the peer review process, students are required to submit initial drafts as early as a week prior to the deadline for the assignment. This gives them significant time to improve the assignment on their own, as well as integrate feedback from at least one peer. In submission of their final draft, students are asked to include a letter to the editor which highlights major revisions that were made based on both peer feedback, as well as faculty feedback from prior assignments. This holds students accountable for both reading and applying faculty feedback as they continue throughout a course. Since introducing the peer review process in the lab, we have seen significant improvement in the quality of communication assignments, especially with regard to minor mistakes that can be time-consuming to address when grading. As a result, grading can be completed more quickly with a focus on higher-level feedback.

564655 Communication Skills Instruction in a Two-Semester Capstone Design Course Sequence

The authors serve as instructors for the Chemical Process Design II (CHEN 4316) and Design III (CHEN 4317) course sequence in chemical engineering at Texas A&M University-Kingsville (TAMUK). An important aspect of these two courses is instruction and mentoring of students in the development of their communication skills. The two principal reasons that communication skills are taught in this course sequence are (a) a necessity to prepare students for a successful engineering career, and (b) to satisfy the ABET student outcome 'g' (new ABET student outcome '3'), which is "an ability to communicate effectively with a range of audiences". Additionally, to meet a SACS (Southern Association of Colleges and Schools) requirement, the Design III course is designated as "writing intensive" for the chemical engineering undergraduate degree.

Prior to this design course sequence, undergraduate chemical engineering majors have completed three courses related to communication skills, specifically two courses in undergraduate English Rhetoric and Composition, and a Communications course in Professional Communication. Upon entering the capstone design sequence, most students indicate they do not yet feel comfortable with their communication skills, in particular their oral communication skills. Therefore, both oral and written communication skills are emphasized throughout both of the courses in this sequence. One group oral presentation is required in Design II, while four group oral presentations are required in Design III. For all of these presentations, individual feedback is provided to the students on their presentation skills. Additionally, the instructors will also go over presentation do's and don'ts as a series of talks occurs over several class days. Three written group project reports are required in Design II. In the Design III course, the writing requirements are greater because of the writing intensive aspect of the course—ten individual writing assignments, along with a group mid-term report and a group final design report. At least two of these written submittals in Design III require draft and final versions, with the students incorporating changes according to instructor comments for the final. This presentation focuses on the successes and challenges associated with the intensive communication instruction in this capstone design course sequence at TAMUK.

566319 Framework for Conducting a Technical Communication Crosswalk across a Chemical Engineering Curriculum

ABET student outcome 4 states that undergraduate students must have "an ability to communicate effectively with a range of audiences" upon graduation. To meet this student outcome, and enhance the communication skills of students, an undergraduate engineering program can intentionally place technical communication instruction across its curriculum. However, not all engineering programs intentionally scaffold experiences to develop technical communication skills through the curriculum. To assess the number, placement, and quality of communication-related graded events within our chemical engineering curriculum, we developed a technical communications crosswalk framework. The framework included mapping of written communication events (research papers, contemporary issue papers, capstone design reports, resume writing exercises, and short technical communications), oral presentations, and laboratory reports from freshman through senior year in a total of 16 required courses for our chemical engineering majors. Linkages for each type of technical communication experience to experiences in course pre-requisites were also included to determine opportunities for connecting events across courses. A comprehensive, multiple-part survey was used to collect characteristics of each technical communication event, to include, the topic, whether the event was completed by individual students or in a team, the percentage of the total course grade the event was worth, and where the graded event was placed within the course, amongst others. Instructor assessment of each event was also collected for courses taught by chemical engineering faculty (e.g., mass and energy balances, chemical reaction engineering, etc.). The technical communications crosswalk framework provided insight into the way our program addresses ABET student outcome 4 and identified some areas for possible improvement. For example, we are considering how to improve technical communications early in the student experience, which could include developing a 1.0 credit technical communications seminar taken in the sophomore year. While the results will vary, the crosswalk framework introduced in this study can be applied to engineering programs at many universities.

564578 Theory, Design, and Early Implementation Results of a Multi-Tiered, Scaffolded Writing Pedagogy

As undergraduate engineering students graduate and advance in their careers, they are faced with multiple tasks that require them to write extensively, whether that be in industry or graduate school. Because of the insufficient writing requirements in undergraduate engineering curricula, many engineers are unprepared for the writing-demand necessary to convey their own ideas or understanding of ideas. More rigorous writing practices would not only improve effective communication skills as undergraduate students pursue their education, but this can also help students develop a deeper conceptual foundation of engineering topics. The writing pedagogy of interest follows a scaffolded approach and was implemented alongside a current hands-on learning pedagogy in a 46-student, junior-level fluid mechanics class (CHE 332) at Washington State University. Students began by taking a pre-test with a descriptive paragraph of the technical phenomena occurring in a Venturi meter; certain phrases of the paragraph were designated as having two options, where they had to choose the correct phrase. Students then used the Venturi hands-on learning device, followed by a posttest. In the posttest, the sections that were previously presented in the form of two options were changed to blanks, where students were required to write their own comprehensible phrases. The results of this implementation were compared to a previous semester when the scaffolded writing pedagogy was not used, with that cohort of students acting as the control group. From pre- to posttest, it was found that the experimental group performed significantly better than the control group with 99% confidence and a medium effect size, based on ranges for partial eta squared. These initial steps are intended to help students learn how to discuss engineering concepts, leading them to properly write full paragraphs on their own on the final exam in early-May. The results of the final exam will be included in the presentation.

571808 Remaining True to Yourself

The sudden transition from being a Postdoctoral Researcher, working under a principal investigator, to becoming an Assistant Professor leading an own group, can be a daunting balancing act, somewhat akin to starting a business, but with the unique human dimension associated to working in education.

There is a wonderful sense of purpose, with vastly increased independence and the opportunity to mentor, but the journey, navigating the complex and ever-evolving space of academia, is also full of challenges: from realizing a personal vision and leadership to the daily operational elements; from preparing and teaching new courses to setting up a lab; from raising funds for research, to tutoring and advising students. And, as the ultimate challenge: Remaining true to yourself and maintaining a healthy balance in life.

I will share some thoughts, based on my international and US-based experiences, from Assistant Professor to Head of Department, via professorships in the Netherlands, the USA and the UK.

566789 Reflections from Ten Years at the Lectern: Insights on Teaching, Scholarship, and Service from an Educational-Focused Perspective

If asked to identify what are the three most important elements of a professor's life, a common reply might be, "teaching, scholarship, and service." As faculty members progress through their careers, it is important to identify how they want to make an impact in these areas. It is understandable that different types of universities place different emphasis on these three elements: an R1 research university might put more emphasis on research, whereas an primarily undergraduate institution (PUI) might put more emphasis on teaching. But what about service? Service can mean many things: service to your department, service to your institution, and service to your profession.

In this talk, I will provide insights into advancing teaching, scholarship, and service from the perspective of working at an undergraduate-focused institution. More specifically, I will focus primarily on teaching and service. The teaching part of my talk will focus on adopting modern pedagogical methods and leveraging resources to enhance student learning in the classroom. Examples of teaching initiatives that I have worked on include peer instruction, digital textbooks, and incorporating computation in the classroom. The service part of my talk will provide insight into how to identify opportunities to make a difference both on an institutional and professional society level. Examples of service initiatives that I have worked on include serving on multiple search committees, chairing my university's Faculty-Student Senate, leading institutional accreditation efforts, serving as a volunteer ABET evaluator, and working in leadership roles at AIChE and ASEE.

While I will provide some background on the opportunities that I have pursued in the areas of teaching, scholarship, and service, ultimately the goal of this talk is to provide advice to young faculty members in achieving success and ideally a valued sense of fulfillment in these three important areas of faculty life.

577504 K-12 Outreach: An Emphasis on Communication

The new ABET student outcomes now include an ability to communicate with a ***range*** of audiences. It has been standard practice that throughout our curriculum, our students primarily practice their communication skills with their peers, faculty members and professional engineers. The attainment of the new outcomes will better prepare our graduates for professional practice when they will be required to communicate with a non-technical audience. Our K-12 outreach project, which is a required part of several of our core courses, allows our students the opportunity to teach technical information to a non-technical audience and determine the effectiveness of their presentation. Details of the outreach project, including K-12 logistics will be shared during this presentation - as well as, examples of hands-on activities, demonstrations, assessment and some of our student misconceptions which have been uncovered.

578154 Increasing Chemical Engineering K-14 Outreach By Tackling Water Conservation Using Human-Centered Design

How much water does the average person use at home per day? The U.S. Geological Survey reports that old showers may use up to 5 gallons of water per minute and water-saving shower heads produce about 2 gallons per minute. The agency recommends that taking a *shorter* shower using a low-flow showerhead saves lots of water. Even if an individual invests in a low-flow showerhead, they still need to figure for how long they are showering. A water consumption sensor that relatively anyone can install in their shower head can help to solve this problem. In addition, global market research reported that the environmental sensor market was valued at USD 1.22 billion in 2017 and is expected to reach USD 2.19 billion by 2023.

The use of sensors in the K-14 classroom has shown to be of significant educational value because they can help address societies problems while integrating many disciplines such as chemistry, physics, mathematics, computer science, engineering, civics, among others. Exposure to the use of sensors to solve real problems can encourage students to follow career paths in Science Technology, Engineering, and Mathematics (STEM), particularly for underrepresented communities. The primary objective of this pilot project is to create a water consumption sensor that is easy to use, and cost-effective for the everyday individual. The integrated sensor system uses simple electrical equipment including photo-resistors, light sources, waterproof cabling, Legos, 3D printed parts, and Arduino. Ideally, this work will become an instructional module that can be used to increase the interest of K-14 students into the chemical engineering field by tackling environmental problems using a human-centered design approach where the needs and wants of the user are paramount for the designer.

576769 Increasing STEM Understanding and Engagement through Engineering Learning Activities

Many engineering students enter their first year of college with minimal hands on experience and little understanding of what engineering majors actually involve. Most K12 educators who do attempt to integrate engineering in their classrooms are not engineers and thus, rely on published lesson plans. There are several databases that teachers can use to find engineering lessons, however, it is important for K12 educators to select from among these activities those that are aligned with the disciplinary concepts they teach. There is a need for guidance in selecting, designing, and adapting engineering activities aligned with curriculum objectives. The goal of this study was to help pre-collegiate students understand what engineering is and how engineering concepts are applied to solve real world problems. To do this, we have related engineering applications to science and math concepts that students learn in their classrooms. With the guidance of middle school teachers to identify important underlying disciplinary concepts, we have created and modified several lesson plans that include hands on engineering-related activities with the intent of improving middle school students' understanding of the underlying disciplinary concepts. Pilot studies of some activities have been perform, and these findings will be used to guide the development of curriculum for an engineering summer camp planned as a week-long outreach program at the University of New Haven for the summer of 2020.

569166 Aiche's K-12 Outreach STEM Module Competition: Results from the First Event and Future Plans

AIChE is holding the first iteration of the K-12 Outreach STEM Module Competition on November 10, 2019 at the Annual Meeting. The competition will share and support the outreach work being conducted by many companies, universities and individuals.

Over 100 regional K-12 students, educators, and parents will be invited to observe and interact with STEM modules in the competition. Modules can address any concept within the range of STEM, but will ideally connect to concepts or applications within chemical engineering. During this two hour event, competitors will showcase their module to K-12 students for the first hour, and present the module to any interested AIChE professional meeting attendees for an additional hour.

Each module should have an experimental component and a corresponding written component. The experiential component will include:

- Connection to at least one concept taught in K-12 STEM courses
- Clearly defined objectives and expectations for student involvement
- Sufficient supply of materials based on estimated participation of 100 students / AIChE members
- Proper safety precautions for all contingencies

The written component of the module should have:

- A list of necessary materials / equipment
- Clear procedure
- Safety analysis, with identification and mitigation

All modules will also be evaluated based on accessibility and cost. All module equipment and supplies should cost less than $100 per module. Modules should be accessible enough for K-12 educators to use in classrooms, and a low overall cost will help to achieve this goal.

All competitors will also develop a handout highlighting the K-12 concepts addressed that attendees will be able to take home and further learn from. All competing modules will be saved into a central digital repository for free access by K-12 educators and AIChE members.

This presentation will discuss the results of the first iteration of the competition, and the plans for the competition at future annual and student AIChE meetings.

569474 Outreach Methods and Modules in Partnership with a Geothermal Research Program

Supporting a departmental outreach team, particularly within a research institution, can often require a significant amount of lobbying and charity from faculty and students. Our K-14 outreach program has found success sustaining itself by moving from being solely funded by the department to supplementation through connections with various research programs. In this talk we will detail how we have developed a partnership with a large Department of Energy geothermal fracturing research project, and used that partnership to cultivate additional outreach support through our state government.

Through this work, we have also developed a variety of effective outreach modules around geothermal energy production. These modules include several inexpensive hands-on STEM activities focusing on topics of permeability, thermodynamics, and turbine energy generation. In the first module, students monitor the pressures needed to pass fluid through simulated rock samples at various flowrates. This module is meant to demonstrate the technical hurdles of passing fluid through the dense type of rock particular to the geothermal reservoir of this work. The second module uses small thermoelectric devices to illustrate the necessity of a temperature *difference* to extract useful work. In this module students use their body heat and ice water to power a light. In the last module, students use common blowers for cooling computer processors as turbines to generate electricity with their breath and also light various LEDs. Both the thermoelectric and blower modules use custom, inexpensive voltmeters to add quantitative activities around power generation in order to scale to student's sophistication and meet additional curricular requirements.

569957 How to Organize and Implement an Outreach Day to Excite, Engage, and Educate Middle School Students about Engineering

Although young children seem to have an inherent desire to explore STEM activities, most students lose interest in pursuing a career in these fields in middle school. Exposing students to STEM experiences in middle school, and even earlier, can have a positive impact on their choice of career in the future. With the growing workforce demand for STEM professionals, it is important to reach children at this critical juncture in their development. One reason for lack of interest is that most students do not have a good understanding of what engineers actually do. Since the petrochemical industry has such a large presence in Louisiana, students who do know a little about engineering lack awareness of the wide variety of disciplines within this large field. Many of these middle school students believe that an engineer can only get a job working in the petrochemical industry, resulting in missed opportunities to recruit promising students with interests in human health and the environment. In an effort to address this need, the authors have developed an outreach day (ENGage LSU) that incorporated hands-on learning activities from a range of engineering disciplines including chemical, biological, mechanical, electrical, and environmental engineering. A hallmark of this outreach day is that all of the activities focused on bioengineering and nanoengineering with applications to clean energy and water, tissue engineering, disease diagnostics, and the development of new materials. The program has been offered three times during the spring semester in 2017, 2018, and 2019. In each year, the number of attendees has increased from ~90 students to ~200 students. To accommodate these numbers the program has increased the number of junior faculty members who give the demonstrations from 11 to 16. The majority (83%) of the middle school students come from traditionally underrepresented minorities so this outreach day is a unique opportunity for them to learn more about different fields of engineering. Using pre- and post-survey we observed a significant increase in (i) the number of students who had the confidence to become an engineer, (ii) the number of students who knew what biomedical engineering was, and (iii) that engineers could help doctors diagnose and treat patients. This presentation will provide details on how the program was designed and improved to maximize the impact on the middle schools and increase their excitement about one-day enrolling in a STEM discipline.

564492 Engaging Middle and High School Students in Hypothesis Generation Using a Citizen-Scientist Network of Air Quality Sensors

Poor air quality affects 90% of the world's population and contributes to 7 million premature deaths annually. Typically, government organizations measure air quality at sparsely distributed measurement stations, equipped with expensive, high-quality instrumentation, and the resulting measurements become available after an hour or two delay. Salt Lake City, Utah periodically experiences some of the worst air quality in the nation, and it has complex meteorology and topography. Consequently, the limited number of stations cannot capture community-level variations in air quality. Our team has deployed a network of approximately 100 low-cost particulate matter (PM) sensors and an accompanying website (www.aqandu.org) for improving the community's understanding of air quality and supporting citizen scientists interested in air quality. This network has generated a rich set of PM measurements, capturing how PM levels evolve over time and space, particularly during pollution events. Building on prior educational outreach and citizen science exercises, we developed an interactive, team-based teaching module using local real-world data. This teaching module's goal is to engage students in generating and testing hypotheses while also encouraging citizen use of real-time air quality data for their own interests, such as exploration, science fair projects, or environmental oversight. We have piloted this module with over 300 students across 6 local high schools in various chemistry, engineering, environmental science, and physics classrooms. Structured around a data analysis exercise with local air quality data, the module helps guide students through creating and testing hypotheses about air quality under various conditions. The module also incorporates fundamental analysis tasks, such as loading and plotting data in a spreadsheet program to build students' familiarity with basic data analysis techniques. Using pre- and post-survey responses, this work seeks to evaluate how a guided, team-based outreach module impacts students' perceptions of outdoor air quality, sense of engagement with a data analysis exercise, ability to generate and test hypotheses, and overall success with incorporating publicly available local data from distributed sensor networks into their curriculum. Finally, this module introduces traditionally underrepresented minorities in STEM to distributed data collection, interpolation, modeling, and visualization concepts used for generating community-scale air quality models. Over half of the visited schools had a minority enrollment exceeding 50% of the student body, and all but two surpassing the state average of 25%.

Preliminary results show this module to be highly engaging, and effective for improving students' awareness of air quality's geospatial and temporal variations during a variety of pollution episodes. Survey results also show this module is effective at introducing hypothesis generation and testing techniques. All classrooms reported wanting to host the module again and having plans to incorporate their local air quality data for future activities and assignments.

Drs. K. Kelly and P.-E. Gaillardon have an interest in the company Tetrad: Sensor Network Solutions, LCC, which commercializes solutions for environmental monitoring.

560040 Strategies for Implementation of Desktop Learning Modules

Hands-on demonstrations are being developed to complement lecture-based courses and clarify common misconceptions in the chemical engineering curriculum, such as friction losses leading to decreases in downstream fluid velocity. An example of these hands-on demonstrations is the desktop learning modules (DLMs) developed by Bernard Van Wie and collaborators at Washington State University [1-5]. While data has shown that integration of these modules into the curriculum improves student learning outcomes, wide-spread adoption can be difficult to achieve. In this study, both fluid mechanics and heat transfer DLMs were implemented across four courses at the University of Kentucky (Lexington and Paducah campuses) to class sizes ranging from 4 to 50 students. To successfully implement these DLMs, strategies were developed on a class-by-class basis to overcome barriers such as large class size, lack of classroom resources and inflexibility in the scheduled curriculum. Strategies include in-class integration for smaller class sizes of 4-30 students vs. out-of-classroom assignments for larger classes (50 students) utilizing a shared departmental laboratory space. Across all courses, pre- and post-tests were used to evaluate the effect of module use on student learning outcomes. Additionally, some courses integrated in-class controls, allowing for comparison of learning outcomes across the spectrum when learning modules or traditional lectures were used to relay information. When compared to a fully-controlled experiment at WSU, similar learning outcomes were achieved for the personalized-implementation plans. For instance, post-test scores were increased by on average 30% and 20% for velocity in a pipe and velocity in a venturi meter, respectively. This compares to increases of approximately 30% and 15% for the WSU study. By developing course specific implementation strategies, hands-on learning modules can be adapted across a wide-range of classroom conditions allowing for wide-scale implementation.

1. Beheshti Pour, N., D.B. Thiessen, R.F. Richards, and B.J. Van Wie, *Ultra-Low-Cost Vacuum Formed Shell and Tube Heat Exchanger Learning Module.* International Journal of Engineering Education, 2017. **33**(2A): p. 723–740.
2. Kaiphanliam, K., N. Beheshti Pour, D. Thiessen, and B. Van Wie, *Desktop learning modules (DLMs) and their effects on student progression through Bloom's taxonomy for fluid mechanics concepts*, in *American Insitute of Chemical Engineers*. 2018: Pittsburgh, PA.
3. Khan, A., N. Beheshti Pour, F. Meng, D. Thiessen, P. Dutta, R. Richards, P. Golter, and B. Van Wie. *Low-cost, transparent, hands-on fluid mechanics and heat-transfer experiments for the classroom*. in *American Physical Society*. 2018. Atlanta, GA: Nov. 13 - 20.
4. Nazempour, A., N. Beheshtipour, S.W. Njau, B.J.V. Wie, J.K. Burgher, P.B. Golter, R.F. Richards, F.S. Meng, O.O. Adesope, C.D. Richards, N. Hunsu, D.B. Thiessen, and A.D. Graviet, *Miniature Low-Cost Desktop Learning Modules for Multi-Disciplinary Engineering Process Applications*, in *Annual Conference of the American Society for Engineering Education*. 2015: Seattle, Washington. p. 1.
5. Meng, F., B.J. Van Wie, D.B. Thiessen, and R.F. Richards, *Design and fabrication of very-low-cost engineering experiments via 3-D printing and vacuum forming.* International Journal of Mechanical Engineering Education, 2018. **Published On-Line**: p. 0-29.

569226 Enhancing the Design of Low-Cost Desktop Learning Module Kits for Ease of Implementation and Adoption

The development of novel, visual, hands-on learning tools to improve student understanding of complex fluid flow and heat transfer concepts is becoming increasingly popular. However, many of these systems are subject to a number of factors that prevent widespread adoption including cost, complexity, and size. Thus, the developers of hands-on learning tools should focus on producing modules which are flexible in application and require minimum preparation time from professors for easy adoption. We have previously developed numerous Low-Cost Desktop Learning Modules (LC-DLMs) including a hydraulic loss system, venturi meter, double pipe heat exchanger, and shell and tube heat exchanger. Although these were effective for enhancing student learning, they required the professor and teaching assistants to spend up to several hours preparing the module kits prior to classroom implementation. To address this, we have recently redesigned the LC-DLM kits to reduce the amount of pre-class preparation time to less than a half hour. With the simplicity of our new kits, students are able to set-up the modules on their own in less than five minutes, and the compactness of the design increases application flexibility by allowing students to fit the LC-DLMs in their backpacks so that they can be used outside the classroom or at home. This presentation will be focused on describing the successful implementation of our new venturi meter and shell and tube heat exchanger LC-DLMs during a junior-level Fluid Mechanics and Heat Transfer course. Preliminary data shows that student performance significantly improved between pre- and post-assessments for several questions within the assessment; post assessments were given immediately after and 2-weeks post-implementation for the heat exchanger and venturi LC-DLMs, respectively. Results from a survey addressing student attitudes towards the new module kits will also be presented. We are confident that our optimized module set-up will allow for easier implementation at diverse institutions by minimizing the complexity of the equipment and the time required in preparing for implementation.

570487 Temperature Control Lab for Dynamics and Control

The 2015 NSF-sponsored report "Chemical Engineering Academia-Industry Alignment: Expectations about New Graduates" [1] identified a strong industrial need for practical understanding of process control and system dynamics. Industry feedback also suggests more focus is needed on translating process control theory into practice [2]. At many universities, this need is met by integrating laboratory experiences into process control courses. However, increased enrollments in chemical engineering strain the available laboratory resources and make these labs difficult to schedule and manage. For these reasons, we have developed small, inexpensive, and portable process control lab designed to reinforce process dynamics and control theory. The Temperature Control Laboratory (TCLab) hardware consists of an Arduino shield that fits onto a standard Arduino UNO or Leonardo microcontroller, a 2A, 5V power supply, and a USB cable for communication with a PC. For reliability and safety, there are no moving parts and there is a safety shut-off to prevent temperatures above 100°C. The TCLab software consists of firmware that can be preloaded on the Arduino microcontroller, an open-source Python library providing a laboratory interface, real-time plotting and data logging services, and a library of Jupyter notebooks demonstrating topics in process control.

From a modeling and identification perspective, TCLab is a simple multivariable system comprised of 2 heaters and 2 closely coupled temperature sensors. The step response of each heater/sensor pair can be approximated by a first order system with a time constant of 2 minutes. The interaction between the heater/sensor pairs is small enough that SISO control is effective, but sufficient to give an introduction to multivariable modeling and control. Thermal energy from the heater is transferred by conduction, convection, and radiation to the surroundings and to the temperature sensor. A detailed energy balance provides a high-fidelity model for each heater/sensor pair that can be extended to a linear state space model with four states, two inputs, and two measured outputs for the entire device. All model parameters can be identified with step test experiments.

The TCLab module can demonstrate many process control modalities. Basic single-input single-output control adjusts heater power output to maintain a temperature setpoint. Interacting two-input, two-output control can be demonstrated by simultaneously controlling both heaters. Cascade control is demonstrated by manipulating one heater and two sensor measurements to control the temperature of the other device. Students can implement the control modalities by coding control algorithms including relay, PID, and model predictive control.

The TCLab device is also a platform for demonstrating state estimation. A linear four-state model for the temperatures of the two heaters and two sensors provides excellent description of the observed dynamics. An observer can be implemented to estimate all states from either or both sensor measurements, and used to demonstrate state feedback, the use of soft sensors to control heater temperature, and sensor fault detection.

The small process control experiment is given to students to reinforce concepts in dynamics and control theory. The lab is integrated at various points to reinforce theory with a practical application for the process dynamics and control course. There are a number of potential pitfalls to avoid and effective strategies to adopt to maximize the learning potential of the lab experience. The lab kit is used to teach digital twin modeling, model identification, and dynamic parameter regression. The objective of this presentation is to give additional details on the learning objectives and discuss the educational outcomes. Lab kits have been distributed to early-adopting universities that include Notre Dame, Iowa State, Oklahoma State, Georgia Tech, New York University, Louisiana Tech, McMaster University, Christian Brothers University, Villanova, University of Iowa, Brigham Young University, University of Pretoria, Western Michigan University, and Zaragoza University. Companies such as APCO, Inc. have developed commercial software for the TCLab into SimTune for training on industrial controllers such as Allen Bradley and Honeywell PID control. Approximately 40 universities use the TCLab for the process dynamic and control course and several faculty have developed new learning modules in a variety of spoken and computer languages. The modules consist of several learning phases: (1) generate a step response, (2) estimate parameters to determine a model, (3) tune a PID controller, (4) reject disturbances and track setpoint, (5) develop SISO and MIMO control methods, and (6) advanced topics such as Model Predictive Control (MPC) [3] and machine learning [4]. This lab was also presented at the 2017 ASEE Summer

School at NCSU as part of resources for teaching process transient analysis. Lab solutions for instructors are provided in Python and MATLAB.
See https://apmonitor.com/heat.htm for additional lab details.

[1] Luo, Y., Westmoreland, P.R., Chemical Engineering Academia-Industry Alignment: Expectations about New Graduates, An NSF-Sponsored Study led by the American Institute of Chemical Engineers, URL: https://bit.ly/2HSnVzd, 2015.
[2] Alford, J. and Edgar, T.F., 2017. Preparing chemical engineering students for industry. Chem. Eng. Prog, 113(11), pp.25-28.
[3] Beal, L.D.R., Hill, D., Martin, R.A., and Hedengren, J.D., GEKKO Optimization Suite, Processes, Volume 6, Number 8, 2018, doi: 10.3390/pr6080106.
[4] Hedengren, J.D., Advanced Temperature Control, URL: https://apmonitor.com/do/index.php/Main/AdvancedTemperatureControl Retrieved: Apr 11, 2019.

567988 Everyone Loves Snack! Classroom Experiments You Can Eat

According to the annual curriculum survey, the modal number of laboratories in the Chemical Engineering curriculum is two. The fact that students benefit from hands-on learning and direct observation of principles across the curriculum has us to develop experiments and demonstrations that can be run during the lecture period, in typical classrooms. In such situations, it's nice to be able to use food as the experimental subject because it is welcome in most classrooms (while chemicals may not be) and, as the title says, everyone loves snack. In this talk, I'll discuss several classroom-friendly experiments adaptable for a variety of core chemical engineering courses, their learning outcomes, and how to run them.

Some of this work was previously shared by the author in the AIChE EdDiv Webinar series.

577708 Reinforcing Design of Experiments, Regression, and Optimization with Paper Helicopters

This presentation will walk through a module developed for our sophomore/junior level numerical methods course for chemical engineering students. In brief, the students are divided into small teams with the objective of creating a paper helicopter with the best loft time. They are given 32 rectangles of paper to conduct their experiments. Three significant geometric variables are identified: rotor length, stem width, and stem length. The students then use the class period (80 min) to design their set of experiments based on response surface designs taught previously. They run their experiments, and then fit their raw data to a quadratic model. They determine if the model can be reduced based on the significance of model coefficients. They then use this model to determine the optimal paper helicopter design. The class ends with each team submitting their best candidate and having a face-off competition with bragging rights and glow stick glasses as the ultimate prize. This course is taught using MATLAB, but could be readily adopted to other software packages.

https://www.cbe.iastate.edu/blog/2017/10/27/teamwork-takes-flight-with-paper-helicopters-in-cbe-class/

578118 A Demonstration Kit for First-Order Processes and Compartmental Models

First order processes are integral to the chemical engineering curriculum from Fick's law to reactor design to pharmacokinetic modeling. Calculations involving these processes often involve differential equations, the solutions of which are sometimes difficult for students to visualize and conceptualize in real time. Here I present a demonstration that uses water, a pump, and dye to simulate first-order processes such as filling a tank, a continuously-stirred tank reactor, a one- or two-compartment pharmacokinetic model, equilibrium reactions, and more. The entire apparatus can be assembled for under $100 and used in courses across the chemical engineering curriculum as a five to ten minute in-class demonstration.

575915 Optimizing National Dissemination and Use of Low-Cost Desktop Learning Modules

Low-Cost Desktop Learning Modules (LC-DLM), developed at Washington State University (WSU), are hands-on modules that are used in conjunction with traditional lectures, such that students can see for themselves the real-world effects of the concepts and theories that they are learning in class. Prior work at WSU has demonstrated that LC-DLMs are effective at increasing students' understanding of the concepts. We have taken steps to spread the use of these LC-DLMs to other universities and have developed a combination of technological tools, as well as a national "hub and spoke" method in order to facilitate the use of these DLMs at other institutions. We have developed a comprehensive webpage that includes online video demonstrations, worksheets, assessments, and frequently asked questions to stimulate implementation of LC-DLMs and minimize potential barriers to adoption. Additionally, hub coordinators have attended the first dissemination workshop at WSU and will take the training and information they received to advance LC-DLM use at their home institutions. At least one workshop will be held each year at a different "hub" institution, where individuals from "spoke" institutions will learn how to use LC-DLMs effectively in their classroom from hub coordinators and representatives from WSU. Lessons learned from workshop coordinators and feedback from participants, such as the need for clear step-by-step instructions and different strategies for developing a control group, will be shared to assist others in developing parallel strategies for propagating the use of new instructional approaches.

567893 Let's Get Pumped: A Student Hand-Pump Project Intcorporated into Fluid Mechanics

The design component of chemical engineering curricula commonly entails a one or two-semester senior capstone project where students simulate a chemical plant. Although this design course enables students to think in terms of sizing equipment and industrial reactions for chemical plants, students often lack the ability to physically design and build equipment.

Here we have integrated a hand-pump design project for students in the fluid mechanics (transport phenomena I) course. Students were broken into teams of three to five and tasked with building a light-weight, cheap and effective hand pump that includes household materials. Students were encouraged to do their own research and learn how pumps operate outside of class. At the conclusion of the project, student teams demonstrated the capabilities of their pump and were tasked with giving a shark tank style pitch to convince the audience why their company should receive an investment. This project provides students with an opportunity to be innovative by building a physical product and to think from an entrepreneurial perspective.

583199 Introduction

A concise overview of the 50th Anniversary of Computer Aides for Chemical Engineering (CACHE) Conference on "The Future of Cyber-Assisted Chemical Engineering Education," held in Breckenridge, Colorado, in July 2019 will be provided.

Dissemination plans for the conference outcomes include publication of several articles in the flagship magazine of AIChE, Chemical Engineering Progress, and a special issue of Chemical Engineering Education.

583200 CACHE History

The proposal founding the Computer Aids for Chemical Engineering Committee was submitted to the NSF on July 14, 1969 and first official meeting of the CACHE organization was held on November 18, 1969 in Washington DC. From its inception, CACHE has promoted the development and distribution of technology-based materials and software in chemical engineering education through projects applying computational chemical engineering, sponsoring conferences, recognizing outstanding contributions, and providing leadership in chemical engineering education. The purpose of this presentation is to review the accomplishments of CACHE over this 50-year period in supporting educational and research innovations.

583201 CACHE Initiatives: Systems Biology

In 2005, CACHE organized a new conference series entitled FOSBE (Foundations of Systems Biology in Engineering). The goal was to create a focused meeting on the topic of systems biology, as it relates to engineering methods and approaches, with a combined emphasis on: (i) applied problems from the pharmaceutical industry, (ii) research methodologies, and (iii) challenges in education and training.

The types of questions that were intended to be addressed by the conference included:
• How can one use systems biology and genomic data to analyze, interpret, and predict the relation between an organism's genotype and its phenotype?
• What are the best methods to combine data from hypothesis-driven research with data from high-throughput studies to create models of cells, communities of cells, and entire organisms? How can one drive iteration and innovation in this model building process?
• What is an appropriate computational infrastructure for maximizing the mining of bioinformatic data? What are the data format and databasing challenges?
• How do the complex network structures constrain intracellular signaling processing?
• What are the promising techniques for estimating model parameters from high throughput data records? How can identification methods be used to drive effective design of experiments?
• How can biological domain knowledge be combined with systems engineering methods to yield model reduction methodologies that capture essential features of biological regulation across multiple scales of time and space?
• How do we educate, train and develop the systems biologists of the future?

In this talk, I will report on the state of the FOSEB conference series, its unique position among other symposia and conferences, its collaborations with other professional societies, and the plans for the future.

583202 CACHE Initiatives: Molecular Modeling

The molecular modeling initiative in CACHE began in 1995 as its first foray into computational-based chemical engineering education beyond process design, operations and control. Molecular modeling has now been firmly established as one of the core competencies of CACHE, with around one quarter of the Board of Trustees representing this effort. In this talk, we will review the activities of CACHE in the field of molecular modeling, which includes educational initiatives and the establishment of the successful Foundations of Molecular Modeling and Simulation (FOMMS) conference series (fomms.org) that has been held every three years, beginning in 2000.

583203 CACHE Initiatives: Process Systems

This presentation describes CACHE initiatives in the area of process systems. CACHE supports the teaching of chemical engineering courses by managing webpages that link to resources that include syllabi, course schedules, computer-aided tools, interactive simulations, screencasts, concept questions, textbook information, useful links, and course notes. The resources relevant to the teaching of process systems include computer-aided educational tools developed by CACHE, such as sustainable manufacturing educational modules.

CACHE organizes conferences on progress and future directions in process systems, which include Chemical Process Control, Foundations of Computer Aided Process Operations, and Foundations of Computer-Aided Process Design. Activities and results are summarized for a newly founded conference on Foundations of Process Analytics and Machine Learning.

583204 Beyond A -> B: Computational Approaches for Education in Reaction Engineering and Kinetics of Complex Systems

Chemical reaction engineering (CRE) is a hallmark of the education of chemical engineers (ChE). It is often cited as a distinguishing feature a ChE compared to other engineers from other disciplines due to its clear marriage of chemistry and engineering. This presentation will focus on cyber-enabled tools that can be easily integrated into the undergraduate curriculum to introduce students to more complex reaction systems governing archetypal chemical processes, where an understanding of reactions well beyond A ® B is essential to guide process design and improvement of both novel and mature technologies. Automated tools for reaction network generation and approaches for automated model building will be discussed. Alternative solution methods for complex reaction systems based on stochastic simulations will be introduced, and software packages that are accessible for undergraduates to carry out kinetic Monte Carlo (kMC) simulations will be outlined. The inclusion of kMC as a core element of a CRE course will be motivated through examples of phenomena that are difficult to capture through analytical rate laws and how intuitive event-based simulations are to students.

583205 Graduate Research & Training in Molecular Data Science Through a University Wide Approach to Integrative Data Science Education

All areas of science and engineering have been the beneficiaries of a data-deluge. Chemical Engineering is at the forefront of this wave as a result of the commoditization of tools such as robotic instrumentation for molecular synthesis and characterization, ever increasing computational resources for simulation and design, high throughput experimental methods including gene sequencers, a proliferation of inexpensive and networked sensors and mixed modality data sources. To be competitive in the changing landscape of industry and academia, the next generation of chemical engineers will need to know how to effectively use data science tools such as statistical and machine learning, scalable data management and visualization techniques to communicate their findings effectively to stakeholders and peers.

The University of Washington (UW) has embraced data science in a campus wide initiative and the Chemical Engineering (ChemE) department at UW has led the effort to integrate highly contextualized molecular data science instruction at the graduate level, for both MS and PhD. ChemE @ UW offers two transcriptable options to graduate students, Advanced Data Science & Data Science. Students who participate in these options get a graduate degree in Chemical Engineering with a transcripted recognition of their participation in the program. The former is designed for chemical engineers who want to design and create new data science algorithms and tools, e.g. new machine learning methods for sparse molecular descriptor spaces, while the latter is intended for data science tool users who want to be confident in their selection of methods, e.g. choosing an effective neural network architecture for a predictive model of blood brain barrier permeability.

This presentation will review the structure of ChemE @ UW's graduate data science options from its inception as a result of two NSF training grants and in the context of the UW campus wide investment in data science education, research and training. It will highlight the NSF NRT Data Intensive Research Enabling Clean Technology (DIRECT) program which has been the primary vehicle for sustainable graduate molecular data science education in the department that includes 1) new graduate coursework – a sequence of flipped-classroom style courses that teach Data Science skills in a manner contextualized to molecules and 2) A capstone project that uses project based learning to practice and apply data science skills to challenging real world problems (in conjunction with external partners) in a team-based setting. It will close with outcomes from early cohorts that participated in the transcriptable options.

583206 Survey: Computing in Chemical Engineering

In 2003 the CACHE (Computer Aids for Chemical Engineering) Corporation carried out a survey of recent graduates in chemical engineering from four universities to determine how that group used (or did not use) computing in performing their job. Since that time, there have been considerable changes in the field of information technology, hence it is appropriate now to assess how the general conclusions reached fifteen years ago have changed. It is also valuable to have such information in order to consider changes in the curriculum, especially with respect to the use of and instruction about computing. Therefore we are conducting a similar survey to the one used in 1997 and 2003.

This survey is in two parts. One part is being emailed to chemical engineering faculty to assess what computing tools are currently being taught and used in the chemical engineering curriculum at universities. A second survey is being emailed to engineers in the chemical industry to assess computing in industry.

570085 Establishing a Learning Assistant (Peer Leader) Training Program

Do you ever feel like there is not enough of you to go around? As course enrollments continue to grow, it can become more difficult to utilize active engagement techniques. If only you had a Learning Assistant (LA)! LAs, also called Peer Leaders, are undergraduate students who facilitate group discussions, guide problem solving, and encourage active engagement in classroom settings. To accomplish this, they typically receive specialized training through a pedagogy course. This session will describe a number of program models (for credit versus for pay), will summarize key research citing the benefits for students and LAs, and will present topical areas for training and development. Finally, we will explore resources available for course content, including building confidence and rapport, utilizing questioning strategies, facilitating group work, and the benefits of self-assessment and reflection.

577207 Student-Created Youtube Videos, a Multimedia Assignment, Improves Engaged Student Learning Outcomes in a Large Engineering Classroom

Starting in the Fall 2014 semester, groups of 4-5 junior Chemical Engineering students were challenged to create informative and captivating YouTube videos of a transport phenomenon highlighting a mass or heat transfer principle that they learned in the ChE 342 classroom or beyond. This innovative ChE342 course project has not only successfully transformed boring deductive classroom teaching into social sharing of engineering principles, but also led to the creation of a YouTube channel (https://www.youtube.com/channel/UCBpw-AyX_Sw8EVsKObSiUVw) "*The Fun of Mass and Heat Transfer*" as a global outreach platform. While we initially demonstrated the effectiveness and statistically significant impact of this teaching innovation on student learning (*Chemical Engineering Education*, 2016 50(3), 186-192), we sought to further investigate how this multimedia assignment affects specific engaged student learning outcomes including teamwork and communication, creativity, self-agency, risk taking and social awareness.

Towards this goal, during the course offering in 2016, 2017 and 2018, students were given the choice between creating a YouTube video and presenting a poster for their project, resulting in two cohorts to elucidate the effect of a multimedia assignment over conventional means on engaged learning outcomes. By conducting focus group studies and pre/post project surveys of these two cohorts, we obtained quantitative and qualitative data suggesting that the multimedia assignment had a statistically significantly greater impact on student creativity and risk taking than the poster presentation. Taken together, the Student-created YouTube Video represents an effective, scalable and sustainable teaching innovation, which not only allows students to showcase their work to a much broader audience but offers unique advantages for improving specific engaged learning outcomes.

571703 A Framework for Developing Concept Questions for Active Learning in Large Classes

We present the process for developing content for the Concept Warehouse (CW), a technology-based tool to facilitate concept-based active learning in mechanical engineering. Concept-based active learning has been shown to improve student performance, especially for traditionally underrepresented students, and to increase academic retention of students in STEM programs. The CW provides three distinct but complementary functions: (a) a content repository, (b) an audience response system to deliver content, and (c) learning analytics that provide data to instructors and researchers. Content in the form of approximately 3,000 ConcepTests, 10 research-based Concept Inventories, and several Inquiry-Based Activities and Virtual Laboratories are now available for ChE. After ~4 years of existence, over 1,000 faculty members at over 150 institutions across the US and internationally are using it.

High quality content is critical for effective concept-based instruction. One of the seven overarching principles for question-design (Koretsky et al., 2014), is the *Emergent Use Principle*: The tool should provide versatility in how questions can be deployed in instruction so that instructors can use them in ways that best fit their beliefs and context. Correspondingly, questions should enable faculty to intersperse formative assessment during more traditional lecture, but also enable those who take a question-cycle (Beatty et al., 2006) approach. In this approach, questions become a central tool of instruction and class time is spent around discussion, arguing, and sense-making. Use of the tool can also include having students write justifications to their selected multiple-choice answer, preparing them to engage in discussion. Thus, the questions are inherently tools of learning through engagement and interaction mediated by the instructor.

Consequently, the nature of effective questions for concept-based active learning can be quite different than those used in formal summative assessments (e.g., concept inventories, standardized tests). We have developed a process for question development that explicitly identifies three goals:

- Content goal: What disciplinary concept or big idea do we want to illuminate?
- Process goal: How must the students use the concept/big idea? What cognitive skills do we want students to exercise?
- Epistemological goal: What ideas about learning and doing engineering do we wish to reinforce?

We present questions with similar content goals but different process and epistemological goals to illustrate the ways question design and intent of instruction interact. While these goals were used for concept-based instruction, we will discuss how they can be applied to other student-centered active learning pedagogies.

Beatty, I. D et al. (2006). *American journal of physics*, **74**(1), 31-39.

Koretsky, M.D et al. (2014). *Advances in Engineering Education*, **4**(1).

573030 Problem Solving When Using Student-Written Problems That Reverse-Engineer Youtube Videos

Problem solving is a desired skill for engineers in the 21st century. In order to promote problem solving amongst undergraduate students and address some issues related to solution manuals, our current study utilized YouTube problems created by previous students by reverse engineering YouTube videos as an alternative to traditional textbook problems. This study replaces traditional textbook problems with YouTube problems and uses a quasi-experimental, treatment/control group design. One focus is to examine the rigor of problem types, and another task is to monitor students' problem-solving when solving textbook and YouTube problems. Data were collected and evaluated using multiple measurement instruments. First, the rigor of problems was examined using the NASA Task Load Index – a 6 question, self-reported survey. Problem solving was assessed using a previously-developed rubric called PROCESS modified to measure: Problem definition, problem Representation, Organization of information, Calculations, solution completion, and Solution accuracy in material and energy balances. The treatment group completed ten textbook problems and nine YouTube problems, while the control group only completed the textbook problems. Results indicate that the rigor of student-written YouTube problems was nearly identical to textbook problems for the same course concepts. While the control group obtained higher PROCESS scores at the beginning of the study, both groups exhibited similar problem-solving skills near the end. Further analysis quantified improvement in problem solving within the treatment group.

566271 Large Classes Create Big Data: Grades, Grading, and Grade Spreadsheets

Teaching over 100 students in a single classroom is a daunting task. Managing grades, grading, and grade spreadsheets for large groups is time consuming and requires skills in student, instructor, and file management. Collecting and handing back homework can be effectively digitized using course management systems, such as Blackboard. In addition, keeping grade spreadsheets organized and up to date to allow for just-in-time teaching interventions is possible. Providing a forgiveness factor can reduce anxiety with course readings and auto-graded homework problems. The latest learnings after many years of teaching large classes will be included.

579815 Approaches to Maintain Student Support on Projects with Increased Class Size

One of the more onerous courses in the chemical engineering curricula is capstone design. This is true of student and instructor alike. This is compounded in a curriculum with a one-semester design sequence as opposed to two. The work presented here is a series of small improvements to save time in properly challenging students with very different skill sets and managing group projects.

Students enter design with varying skillsets depending on their curricular and noncurricular experiences. With this, a competency-based assessment is utilized for topics in the course such as simulation, economics, and safety. The approach allows the instructor and student to agree upon the method of delivery and degree of competency shown prior to the demonstration. This technique appropriately challenges advanced students while providing students with agency over their learning.

Course procedures must be carefully designed to provide peer-to-peer support, improving feedback while also decreasing instructor time giving feedback. The method of project development, peer-to-peer midsemester presentations, utilization of discussion sections, and checkpoints with the instructor are described.

574245 Specifications Grading: Simplify and Save Time

Arguably one of the most tedious and time consuming responsibilities of faculty members, especially in large enrollment courses, is grading. Specifications grading has been touted as an effective method to not only improve grading efficiency and timeliness of feedback, but also student motivation and ownership of learning (Nilson, 2014). This simple grading approach is based on the principle of pass/fail grades with assessments linked to learning outcomes grouped within those required to earn a certain grade. Thus, students can choose from the very beginning of a course which assessments they want to complete in order to earn a particular grade. To earn higher final grades, students must complete assessments that demonstrate higher levels of content mastery. This grading method requires instructors to provide transparent specifications up front, and students must complete work at a minimum standard of B- to earn credit for assessments. They are typically given a token that allows them the opportunity to revise an assignment or submit late work.

The specifications grading approach was implemented in two semesters of a large undergraduate research course and a co-instructed chemical engineering senior laboratory course, both had a variety of individual and group assessments. The grading scheme was explained to the students on the first day of both courses along with a due date matrix for assessments. From the instructors' perspectives, specifications grading saved a significant amount of time compared to traditional grading approaches. The student evaluations of teaching after the first semester indicated mixed feelings regarding the grading scheme. Based on this feedback, the instructors made the motivation for using specifications grading more transparent and provided handouts with checklists to the students during the second semester offering of the course. Details of the implementation of this grading scheme in the laboratory course and in a large research group will be discussed as well as qualitative student survey responses and suggestions for future improvements.

Nilson, Linda B. *Specifications Grading: Restoring Rigor, Motivating Students, and Saving Faculty Time.* Sterling, Virginia: Stylus Publishing, LLC, 2014.

AUTHOR INDEX

AUTHOR INDEX

AUTHOR INDEX

AIChE
120 Wall Street, FL 23
New York, NY 10005-4020

ISBN 978-1-7138-0528-1

CURRAN ASSOCIATES INC.
proceedings.com